Cambridge Tracts in Mathematics and Mathematical Physics

No. 27

MODULAR INVARIANTS

MODULAR INVARIANTS

BY

D. E. RUTHERFORD, B.Sc.

CAMBRIDGE
AT THE UNIVERSITY PRESS
1932

CAMBRIDGE
UNIVERSITY PRESS

University Printing House, Cambridge CB2 8BS, United Kingdom

Cambridge University Press is part of the University of Cambridge.

It furthers the University's mission by disseminating knowledge in the pursuit of education, learning and research at the highest international levels of excellence.

www.cambridge.org
Information on this title: www.cambridge.org/9781107493766

First published 1932
Re-issued 2015

A catalogue record for this publication is available from the British Library

ISBN 978-1-107-49376-6 Paperback

PREFACE

IN the winter of 1929 Professor Weitzenböck pointed out to me that there was no complete account of the theory of modular invariants embodying the work of Dickson, Glenn and Hazlett. The sole source of information on this subject was a number of papers, most of which appeared in American periodicals, and a tract by Dickson which contained the substance of his *Madison Colloquium Lectures*. This tract, while giving a good account of the subject as it was understood in 1914, was published before the modular symbolical theory was instituted. Although the symbolical theory is not yet complete, it certainly affords a much better introduction to the subject than did the earlier non-symbolical methods. The theory is much hampered by the lack of two theorems which seem to be true but for which, as yet, no proof has been given. These are (i) that all congruent covariants can be represented symbolically; (ii) that Miss Sanderson's theorem can be applied to covariants as well as to invariants.

In preparing the present account, the chief difficulty has been the lack of any systematic method of approach, since most of the papers on the subject have been concerned with particular cases only. My aim has been to give a clear and concise account of the theory rather than to give a complete survey of the subject, and I have therefore included in this tract only those methods which seem to be of general application. For the sake of completeness it has been necessary to include the intricate proof of Dickson's theorem in paragraph 13. It is suggested that this might be omitted at a first reading. In order to avoid confusion the reader should notice that the words *fundamental* and *modular* vary somewhat in meaning in the different papers on the subject.

I have, of course, benefited considerably from the papers of Dickson, Glenn, Hazlett, Sanderson and others, and many theorems are taken directly from their papers. The substance of Part II is largely taken from a course of lectures entitled "Algebraische theorie der lichamen" which Professor Weitzenböck delivered in Amsterdam University during the session 1929–30. I have also made use of his lecture notes which he has kindly placed at my disposal. Professor Weitzenböck has been of great assistance to me throughout my work and has given me much helpful advice. My grateful thanks are due to Professor Turnbull of St Andrews University and to Professor Weitzenböck for reading the proof-sheets and for making many suggestions and corrections.

Many thanks are also due to the Syndics of the Cambridge University Press for their helpful criticism of the manuscript.

D. E. R.

St Andrews
April 1932

CONTENTS

PART I

§ 1. A new Notation.

It has been found convenient in this book to introduce the signs $\|$ and $|||$ which shall presently be explained. We say that two numbers a and b are congruent modulo p if their difference is divisible by p. This is commonly written in text-books on the theory of numbers as

$$a \equiv b \pmod{p} \quad \text{......................(1·1)}.$$

The sign $\equiv$, however, often means "is identically equal to," and confusion will arise if we wish to use it with these two different meanings. When it is used as in (1·1) it gives no indication as to whether the congruence holds for all values of a and b or only when a and b belong to a particular field, e.g. x^p is congruent to x modulo p if, and only if, x is an integer (a positive integer if p is even). We shall therefore use $|||$ to mean "is identically congruent to" and $\|$ to mean "is residually congruent to by Fermat's Theorem." Hence the sign $\|$ can only be used in the cases where Fermat's Theorem and its extensions hold. Thus

$=$ means "is equal to,"	e.g.	$x = 4$,	$2 \neq 3$,
$\equiv$ means "is identically equal to,"	e.g.	$4 \equiv \frac{8}{2}$,	$x \not\equiv 4$,
$\|$ means "is residually congruent to," if a is a positive integer,	e.g.	$a^p \parallel a$	mod p,
$|||$ means "is identically congruent to,"	e.g.	$6 \;|||\; 3$	mod 3,
		$a^p \not{|||}\, a$	mod p.

§ 2. Galois Fields and Fermat's Theorem.

A. Speiser* gives the following definition of a Galois Field.—A system of a finite number of elements forms a Galois Field if the following conditions are satisfied:

(i) The elements form a commutative group with respect to the addition law.

(ii) The elements with the exception of zero form a commutative group with respect to the multiplication law.

(iii) For any four elements the distributive law is valid.

From these three conditions all the properties of a Galois Field can be obtained, but it is easy to see that the Galois imaginaries, hereafter

* *Theorie der Gruppen* (second edition), p. 54.

defined, satisfy these three conditions and are therefore the elements of a Galois Field.

Let us consider a polynomial

$$f = a_n x^n + \ldots + a_0,$$

where each of the coefficients $a_n, \ldots, a_0$ is one of the following integers $0, 1, 2, \ldots, p-1$, where p is a given prime, i.e. the coefficients are the positive integral residues modulo p. We may of course suppose that $a_n \not\equiv 0 \bmod p$, and we shall further suppose that f is irreducible modulo p, i.e. that there exist no polynomials ϕ_1 and ϕ_2 of degree less than n such that $\phi_1 \phi_2 \equiv f \bmod p$.

Now it is always possible to find a number b_n such that $b_n a_n \equiv 1$ provided that $a_n \not\equiv 0$, therefore $b_n f \equiv x^n + c_{n-1}x^{n-1} + \ldots + c_0$, where $c_i \equiv b_n a_i$, each c_i again being a residue modulo p of a positive integer. $x^n + c_{n-1}x^{n-1} + \ldots + c_0$ is called the normalised form of f; and obviously $b_n f \equiv 0$ if $f \equiv 0$. Since we are dealing with the case where $f \equiv 0$, we can without lack of generality take the coefficient of x^n congruent to 1.

If $f \equiv 0$, any polynomial in x with integer coefficients is congruent to a polynomial of degree $\leqslant n-1$. Thus for any $\phi(x)$ we get for moduli f and p a residue $\theta(x)$. These residues are the **GALOIS IMAGINARIES** of order n and it is easily seen that they satisfy the conditions (i), (ii) and (iii). Each $\phi(x)$ can be written congruent to some $\theta(x)$,

$$\theta(x) = d_{n-1}x^{n-1} + \ldots + d_0.$$

Since each d_i can be chosen in p different ways, there are p^n different residues and so the **ORDER** of the Galois Field is p^n. Hereafter we shall write this briefly as $GF[p^n]$ while we shall denote the field of all complex numbers by CF.

Fermat's well-known theorem states that if y is a positive integer, then $y^p \equiv y \bmod p$. The generalisation of this theorem for the case where y is a Galois imaginary is obtained as follows. Let u be a Galois imaginary and not zero, then $\ldots, u^{-2}, u^{-1}, u^0, u^1, u^2, u^3, \ldots$ are not all distinct: hence if $u^s \equiv u^t$ then $u^{s-t} \equiv 1$. Let e be the least value of $s-t$ for which this is true, then $1, u, u^2, \ldots, u^{e-1}$ are distinct. So are $u_1, u_1 u, u_1 u^2, \ldots, u_1 u^{e-1}$, where u_1 is not one of $1, u, \ldots, u^{e-1}$. Proceeding in this manner we see that since there are $p^n - 1$ non-zero Galois imaginaries, e must be a divisor of $q = p^n - 1$. Since $u^e \equiv 1$, therefore $u^q \equiv 1$, and if y be any Galois imaginary then $y^{p^n} \equiv y$. In Part I we shall write

$$p^n - 1 \equiv q, \qquad p^n(p^n - 1) \equiv d.$$

The p^n solutions of $x^{p^n} - x \equiv 0$ are the p^n Galois imaginaries, therefore

$x^{p^n} - x \parallel \prod_{i=0}^{q} (x - u_i)$, where $u_0 = 0, u_1, \ldots, u_q$ are the elements of $GF[p^n]$. Further generalisations of Fermat's Theorem were given by E. H. Moore.

For a first reading we would advise the reader to take $n=1$ throughout, thus $q=p-1$ and $d=p^2-p$. The elements of $GF[p]$ are the residues modulo p of the positive integers.

§ 3. Transformations in the Galois Fields.

In the work which follows we shall have to consider groups of homogeneous linear transformations whose determinants do not vanish in the field. The inverse transformations therefore exist. We shall call G the group where the coefficients belong to the CF, and we shall call Γ the group where the coefficients of the transformations belong to $GF[p^n]$. G_1 and Γ_1 are the sub-groups of G and Γ respectively which consist only of transformations whose determinant is equal to unity in the field. When necessary we shall denote the number of variables in the transformation by an upper suffix m, e.g. G^m, $G_1{}^m$, Γ^m, $\Gamma_1{}^m$.

We obtain the order of the group Γ as follows*: Consider the transformations of the following type where the coefficients are elements of the $GF[p^n]$,

$$x_1 = \mathfrak{s}_{11}\bar{x}_1 + \ldots + \mathfrak{s}_{1m}\bar{x}_m,$$
$$\ldots\ldots\ldots\ldots\ldots\ldots\ldots\ldots$$
$$x_m = \mathfrak{s}_{m1}\bar{x}_1 + \ldots + \mathfrak{s}_{mm}\bar{x}_m.$$

There are $p^{nm}-1$ possibilities for the right side of the first equation, for we cannot set $x_1 = 0$. If S and T be two transformations which replace x_1 by the same linear function of the $\bar{x}$'s, then ST^{-1} will leave x_1 unchanged and therefore will have a matrix of the following type:

$$\begin{bmatrix} 1 & 0 & 0 & \ldots & 0 \\ \mathfrak{a}_2 & \mathfrak{b}_{22} & \mathfrak{b}_{23} & \ldots & \mathfrak{b}_{2m} \\ \mathfrak{a}_3 & \mathfrak{b}_{32} & \mathfrak{b}_{33} & \ldots & \mathfrak{b}_{3m} \\ \ldots & \ldots & \ldots & \ldots & \ldots \\ \mathfrak{a}_m & \mathfrak{b}_{m2} & \mathfrak{b}_{m3} & \ldots & \mathfrak{b}_{mm} \end{bmatrix}.$$

Such substitutions form a sub-group of Γ: and if we give the $\mathfrak{b}$'s fixed values there are $p^{n(m-1)}$ possibilities for the $\mathfrak{a}$'s. Also if the $\mathfrak{a}$'s are fixed we have a sub-group of the sub-group the order of which is $O(m-1)$, where $O(m)$ means the order of the group Γ^m; thus

$$O(m) = p^{n(m-1)}(p^{nm}-1)\,O(m-1),$$

and using this as a reduction formula, we have

$$O(m) = (p^{nm}-1)(p^{nm}-p^n)(p^{nm}-p^{2n})\ldots(p^{nm}-p^{n(m-1)}).$$

* Speiser, *Theorie der Gruppen* (second edition), p. 219.

§ 4. Types of Concomitants.

Let $\phi(a, x) = a_{11\ldots1}x_1^l + a_{11\ldots12}x_1^{l-1}x_2 + \ldots + a_{mm\ldots m}x_m^l$ be an m-ary l-ic which is homogeneous in the x's. We shall call ϕ the GROUND FORM. So far as is conveniently possible we shall reserve the letters a, b, c, ... for the coefficients of the ground forms. Consider now the group of all homogeneous linear transformations

$$x_i = \sum_{j=1}^{m} \mathfrak{a}_{ij}\bar{x}_j \qquad (i=1, 2, \ldots, m)\ldots(4\cdot1),$$

subject to the condition that $|\mathfrak{a}_{ij}|$ does not vanish in the field F. Let the matrix of the transformation (4·1) be $\mathfrak{A}$. We shall use small old English text throughout for the coefficients of the transformations. Now let $\phi(a, x)$ become $\phi(\bar{a}, \bar{x})$ under a transformation of the group; then the $\bar{a}$'s will be functions of the a's and the $\mathfrak{a}$'s. Now if any function C of the a's and x's exist such that

$$C(\bar{a}, \bar{x}) \text{ is equal to } M(\mathfrak{a})\, C(a, x) \text{ in the field,}$$

where M is some function of the $\mathfrak{a}$'s only, then C is said to be a CONCOMITANT. If C is a function of both the a's and the x's, it is called a COVARIANT; if it is a function of the a's only, it is called an INVARIANT; if it is a function of the x's only, it is called a UNIVERSAL COVARIANT or an INVARIANT OF THE GROUP. These definitions can readily be extended to the cases where we have more than one ground form. We shall commonly use the term covariant to include invariants.

The a's and the $\mathfrak{a}$'s may belong either to CF or $GF[p^n]$, so that we have the following types:

Type I. If both the a's and the $\mathfrak{a}$'s belong to CF and reductions of the form $p \equiv 0$ are forbidden, the concomitants are then called ALGEBRAIC and this is the classic type treated thoroughly elsewhere.

Type II. If both the a's and the $\mathfrak{a}$'s belong to CF but $p \equiv 0$ is allowed, then we speak about CONGRUENT concomitants. Miss Hazlett* has treated this case as a special case of Type III.

Type III. If the a's belong to the CF and the $\mathfrak{a}$'s belong to the $GF[p^n]$, then the reductions $p \equiv 0$ and $\mathfrak{a}^{p^n} \equiv \mathfrak{a}$ are permitted. In this case we talk of FORMAL concomitants. These have been treated by Dickson †, Sanderson ‡ and Hazlett*.

Type IV. If the a's belong to the $GF[p^n]$ and the $\mathfrak{a}$'s to CF, then

* *Trans. Amer. Math. Soc.* vol. 24, pp. 286–311 (1922).

† *Madison Colloquium Lectures* and *Trans. Amer. Math. Soc.* vol. 15, pp. 497–503 (1914).

‡ *Trans. Amer. Math. Soc.* vol. 14, pp. 489–500 (1913).

$p \equiv 0$ and $a^{p^n} \equiv a$ are allowed. This type has not been treated so far, but we shall give to them the name of NON-FORMAL concomitants.

Type V. If both the a's and the $\mathfrak{a}$'s belong to the $GF[p^n]$, then three types of reductions are allowed, $p \equiv 0$, $a^{p^n} \equiv a$, $\mathfrak{a}^{p^n} \equiv \mathfrak{a}$. We shall call concomitants of this sort RESIDUAL concomitants, but in other papers on the subject they are termed modular concomitants. We prefer to use this term to cover all types where $p \equiv 0$ is allowed. Type V has been treated extensively by Dickson*.

We shall call concomitants of types II, III, IV and V MODULAR to distinguish them from Type I, the algebraic or projective case. By considering the reductions employed it is obvious that every algebraic covariant is also a congruent covariant; that every congruent covariant is also a formal covariant; that every formal covariant is also a residual covariant. It is clear, however, that two formal covariants may be identical when considered as residual covariants or that a formal covariant may be zero when regarded as a residual covariant, e.g. $a_1^2 a_2 - a_1 a_2^2$ is known to be a formal invariant of $f = a_1 x_1 + a_2 x_2$ if $p = 2$. When this is considered as a residual covariant we can use Fermat's Theorem on the a's and our invariant is $a_1 a_2 - a_1 a_2 = 0$.

We can extend our definitions to include what are known as cogredient points. Following Dickson we use the term point in the sense of homogeneous coordinates; thus the point $(y_1 y_2 y_3)$ is identical with the point $(ky_1 ky_2 ky_3)$ and the point $(0, 0, 0)$ is excluded. Now if the transformations (4·1) of the variables $x_1, \ldots, x_m$ and the transformations of the coordinates $y_1, \ldots, y_m$ have the same matrix $\mathfrak{A}$, then $(y_1, \ldots, y_m)$ is called a COGREDIENT POINT. We shall now prove the theorem which will be required later, that *all points whose coordinates belong to* $GF[p^n]$ *are conjugate under* Γ_1, i.e. any such point can be transformed into any other by transformations of Γ_1.

Now $(1, 0, 0, \ldots, 0)$ is conjugate with $(1, a_2, \ldots, a_m)$ under

$$y_1 = \bar{y}_1,$$
$$y_i = a_i \bar{y}_1 + \bar{y}_i \qquad (i \neq 1),$$

and the elements of $(1, a_2, \ldots, a_m)$ can be rearranged with perhaps changes of sign under

$$y_i = \bar{y}_j,$$
$$y_j = -\bar{y}_i,$$
$$y_k = \bar{y}_k \qquad (k \neq i, \; k \neq j).$$

This proves the theorem, e.g. $(2, 1, 0)$ is conjugate with $(1, -2, 0)$ which in turn is conjugate with $(1, 0, 0)$.

* *Amer. Journ. of Maths.* vol. 31, pp. 337–354 (1909) and other papers.

§ 5. Systems and Finiteness.

We shall call a system of covariants $K_0, K_1, \ldots, K_{\sigma-1}$ a FULL SYSTEM if every other covariant I can be expressed in terms of these K_i's. If the σ covariants $K_0, \ldots, K_{\sigma-1}$ are linearly homogeneously independent and if every covariant I can be expressed as a linear homogeneous function of these K_i's, then we shall call the set $K_0, \ldots, K_{\sigma-1}$ a FUNDAMENTAL SYSTEM. The reader should be careful to note that in some papers no distinction is made between a full system and a fundamental system. It is obvious from the above definitions that every fundamental system is also a full system but that not every full system is a fundamental system. As a special case of a full system we have a SMALLEST FULL SYSTEM. The s covariants $N_1, \ldots, N_s$ form a smallest full system if (i) they form a full system; (ii) no full system exists with less than s covariants. These elements $N_1, \ldots, N_s$ form a BASIS of the smallest full system and $K_0, \ldots, K_{\sigma-1}$ form a basis of the full system or the fundamental system as the case may be.

The covariants of a ground form with respect to a group of transformations are said to possess the FINITENESS PROPERTY if there exists a finite full system. We shall not have to prove the finiteness property for any given case, as we shall prove that it holds in every case with which we deal.

That algebraic covariants possess the finiteness property is well known. In another section of this book we shall give E. Noether's proof that modular covariants of all types possess the finiteness property. L. E. Dickson* also gave a proof of the finiteness of residual covariants. We shall consider these finiteness theorems as proved, although we leave the proof till later. Most of the work done in the theory of modular covariants has been concerned with the finding of a basis of a full system in some particular case. As we shall show, there exist several methods of obtaining covariants, but except in the case of residual invariants it is very difficult to say whether a given system is a full one or not.

§ 6. Symbolical Notation.

One of the greatest difficulties for many years in the theory of modular invariants was that no suitable symbolic method of treatment had been found. The method employed in the algebraic invariant theory obviously would not do, since it employed to a large extent multinomial coefficients which in certain cases might be congruent to zero modulo p. Let us take as an example the binary cubic. In the algebraic case we represent it symbolically as a_x^3, where $a_x = a_1x_1 + a_2x_2$.

* *Trans. Amer. Math. Soc.* vol. 14, pp. 229–310 (1913).

Now $$\alpha_x^3 = \alpha_1^3 x_1^3 + 3\alpha_1^2\alpha_2 x_1^2 x_2 + 3\alpha_1\alpha_2^2 x_1 x_2^2 + \alpha_2^3 x_2^3,$$
$$\equiv \alpha_1^3 x_1^3 + \alpha_2^3 x_2^3 \mod 3.$$

Thus for the modular case with $p = 3$, α_x^3 does not represent the general cubic. The method employed in the modular case is merely a generalisation of this one, and it has the advantage that it can be used for the algebraic case also. In general then, we write the ground form as a product of linear homogeneous symbolical factors*. We shall write for example the general binary cubic as

$$\begin{aligned} f = ax_1^3 + bx_1^2 x_2 + cx_1 x_2^2 + dx_2^3 &= \alpha_x \beta_x \gamma_x \\ &= (\alpha_1 x_1 + \alpha_2 x_2)(\beta_1 x_1 + \beta_2 x_2)(\gamma_1 x_1 + \gamma_2 x_2) \\ &= \alpha_1\beta_1\gamma_1 x_1^3 + (\alpha_1\beta_1\gamma_2 + \alpha_1\beta_2\gamma_1 + \alpha_2\beta_1\gamma_1) x_1^2 x_2 \\ &\quad + (\alpha_1\beta_2\gamma_2 + \alpha_2\beta_1\gamma_2 + \alpha_2\beta_2\gamma_1) x_1 x_2^2 + \alpha_2\beta_2\gamma_2 x_2^3. \end{aligned}$$

We shall note that any non-symbolic coefficient of a ground form when represented symbolically must be of the first degree in the α's, first degree in the β's and so on. It is also symmetrical in the symbols $\alpha, \beta, \ldots$. Thus we have the important conditions which a function of the non-symbolic coefficients satisfy:

(i) It must be symmetrical in the symbols $\alpha, \beta, \ldots$.

(ii) Each term of the function must be of the same degree in each of the symbols $\alpha, \beta, \ldots$.

We have other conditions that this function be also a modular concomitant. These will be given in § 9.

We shall use the small Greek letters $\alpha, \beta, \ldots, \delta$ for symbols.

It should be noticed also that with this symbolism no equivalent symbols are required. To illustrate the difference in the two symbolisms we shall consider the discriminant of the binary quadratic.

Let $$f = ax_1^2 + bx_1 x_2 + cx_2^2,$$
$$(\alpha\beta)^2 = \alpha_1^2\beta_2^2 - 2\alpha_1\alpha_2\beta_1\beta_2 + \alpha_2^2\beta_1^2.$$

If we represent f as $\alpha_x^2 = \beta_x^2$, then $a = \alpha_1^2 = \beta_1^2$, $b = 2\alpha_1\alpha_2 = 2\beta_1\beta_2$, $c = \alpha_2^2 = \beta_2^2$ and $(\alpha\beta)^2 = 2ac - \frac{b^2}{2} = \frac{1}{2}(4ac - b^2)$.

If we represent f as $\alpha_x\beta_x$, then $a = \alpha_1\beta_1$, $b = \alpha_1\beta_2 + \alpha_2\beta_1$, $c = \alpha_2\beta_2$ and $(\alpha\beta)^2 = b^2 - 4ac$.

The invariant obtained is the same in both cases, but in the former case it is multiplied by a constant $= -\frac{1}{2}$. An invariant need not, however, have the same form in both symbolisms.

* Sanderson, *Trans. Amer. Math. Soc.* vol. 14, p. 496 (1913).

§ 7. Generators of Linear Transformations.

Capelli* proved that any linear transformations of m variables can be obtained by the successive application of a finite number of linear transformations of the following types:

$$S_h^{(\mathfrak{t})}:\ x_1=\bar{x}_1,\ \ldots,\ x_{h-1}=\bar{x}_{h-1},\ x_h=\mathfrak{t}\bar{x}_h,\qquad x_{h+1}=\bar{x}_{h+1},\ \ldots,\ x_m=\bar{x}_m,$$

$$S_{hk}:\ x_1=\bar{x}_1,\ \ldots,\ x_{h-1}=\bar{x}_{h-1},\ x_h=\bar{x}_h+\bar{x}_k,\ x_{h+1}=\bar{x}_{h+1},\ \ldots,\ x_m=\bar{x}_m.$$

It is easy to see that these in turn can always be generated from $S_1^{(\mathfrak{t})}$, $S_{1,2}$ supplemented by transformations of the type

$$x_1=\bar{x}_j,\quad x_j=\bar{x}_1,\quad x_k=\bar{x}_k\qquad (k\neq 1,\ k\neq j).$$

We have therefore the theorem: *Any homogeneous linear transformation whose determinant is not zero can be generated from transformations of the following three types:*

Type I. $x_1=\bar{x}_1+\bar{x}_2$, $x_i=\bar{x}_i\ (i\neq 1)$

$$\begin{bmatrix} 1 & 1 & \cdot & \cdot & \cdot \\ \cdot & 1 & \cdot & \cdot & \cdot \\ \cdot & \cdot & 1 & \cdot & \cdot \\ \cdot & \cdot & \cdot & 1 & \cdot \\ & & & & \ddots \\ \cdot & \cdot & \cdot & \cdot & 1 \end{bmatrix}.$$

Type II. $x_1=\mathfrak{k}\bar{x}_1$, $x_i=\bar{x}_i\ (i\neq 1)$

$$\begin{bmatrix} \mathfrak{k} & \cdot & \cdot & \cdot \\ \cdot & 1 & \cdot & \cdot \\ \cdot & \cdot & 1 & \cdot \\ & & & \ddots \\ \cdot & \cdot & \cdot & 1 \end{bmatrix}.$$

Type III. $x_1=\bar{x}_j$, $x_j=\bar{x}_1$, $x_k=\bar{x}_k\ (k\neq j,\ k\neq 1)$ e.g.

$$\begin{bmatrix} \cdot & 1 & \cdot & \cdot & \cdot \\ 1 & \cdot & \cdot & \cdot & \cdot \\ \cdot & \cdot & 1 & \cdot & \cdot \\ \cdot & \cdot & \cdot & 1 & \cdot \\ & & & & \ddots \\ \cdot & \cdot & \cdot & \cdot & 1 \end{bmatrix}$$

for the case where $j=2$.

By a combination of matrices of Type III we can interchange any pair of variables thus:

$$\begin{bmatrix} \cdot & 1 & \cdot \\ 1 & \cdot & \cdot \\ \cdot & \cdot & 1 \end{bmatrix}\times\begin{bmatrix} \cdot & \cdot & 1 \\ \cdot & 1 & \cdot \\ 1 & \cdot & \cdot \end{bmatrix}\times\begin{bmatrix} \cdot & 1 & \cdot \\ 1 & \cdot & \cdot \\ \cdot & \cdot & 1 \end{bmatrix}=\begin{bmatrix} 1 & \cdot & \cdot \\ \cdot & \cdot & 1 \\ \cdot & 1 & \cdot \end{bmatrix};$$

* *Lezioni*, p. 202.

by interchanging the variables of Type II we get Capelli's type $S_h^{(\mathfrak{k})}$, and by interchanging the variables of Type I we get Capelli's type $S_{h,k}$.

We shall give a short proof of the theorem in the matrix notation for $m=3$:

$$\begin{bmatrix} \cdot & 1 & \cdot \\ 1 & \cdot & \cdot \\ \cdot & \cdot & 1 \end{bmatrix} \times \begin{bmatrix} \mathfrak{k} & \cdot & \cdot \\ \cdot & 1 & \cdot \\ \cdot & \cdot & 1 \end{bmatrix} \times \begin{bmatrix} \cdot & 1 & \cdot \\ 1 & \cdot & \cdot \\ \cdot & \cdot & 1 \end{bmatrix} = \begin{bmatrix} 1 & \cdot & \cdot \\ \cdot & \mathfrak{k} & \cdot \\ \cdot & \cdot & 1 \end{bmatrix}$$

and

$$\begin{bmatrix} 1 & \cdot & \cdot \\ \cdot & \mathfrak{k}^{-1} & \cdot \\ \cdot & \cdot & 1 \end{bmatrix} \times \begin{bmatrix} 1 & 1 & \cdot \\ \cdot & 1 & \cdot \\ \cdot & \cdot & 1 \end{bmatrix} \times \begin{bmatrix} 1 & \cdot & \cdot \\ \cdot & \mathfrak{k} & \cdot \\ \cdot & \cdot & 1 \end{bmatrix} = \begin{bmatrix} 1 & \mathfrak{k} & \cdot \\ \cdot & 1 & \cdot \\ \cdot & \cdot & 1 \end{bmatrix}.$$

Similarly by use of Type III we can obtain the type

$$\begin{bmatrix} 1 & \cdot & \mathfrak{k} \\ \cdot & 1 & \cdot \\ \cdot & \cdot & 1 \end{bmatrix}.$$

Now

$$\begin{bmatrix} 1 & \mathfrak{a} & \cdot \\ \cdot & 1 & \cdot \\ \cdot & \cdot & 1 \end{bmatrix} \times \begin{bmatrix} 1 & \cdot & \mathfrak{b} \\ \cdot & 1 & \cdot \\ \cdot & \cdot & 1 \end{bmatrix} \times \begin{bmatrix} \mathfrak{c} & \cdot & \cdot \\ \cdot & 1 & \cdot \\ \cdot & \cdot & 1 \end{bmatrix} = \begin{bmatrix} \mathfrak{c} & \mathfrak{a} & \mathfrak{b} \\ \cdot & 1 & \cdot \\ \cdot & \cdot & 1 \end{bmatrix},$$

and

$$\begin{bmatrix} \mathfrak{a}_1 & \mathfrak{a}_2 & \mathfrak{a}_3 \\ \cdot & 1 & \cdot \\ \cdot & \cdot & 1 \end{bmatrix} \times \begin{bmatrix} \cdot & 1 & \cdot \\ 1 & \cdot & \cdot \\ \cdot & \cdot & 1 \end{bmatrix} \times \begin{bmatrix} \mathfrak{b}_1 & \mathfrak{b}_2 & \mathfrak{b}_3 \\ \cdot & 1 & \cdot \\ \cdot & \cdot & 1 \end{bmatrix} \times \begin{bmatrix} \cdot & \cdot & 1 \\ \cdot & 1 & \cdot \\ 1 & \cdot & \cdot \end{bmatrix} \times \begin{bmatrix} \mathfrak{c}_1 & \mathfrak{c}_2 & \mathfrak{c}_3 \\ \cdot & 1 & \cdot \\ \cdot & \cdot & 1 \end{bmatrix}$$

$$= \begin{bmatrix} \mathfrak{d}_{11} & \mathfrak{d}_{12} & \mathfrak{d}_{13} \\ \mathfrak{d}_{21} & \mathfrak{d}_{22} & \mathfrak{d}_{23} \\ \mathfrak{d}_{31} & \mathfrak{d}_{32} & \mathfrak{d}_{33} \end{bmatrix} = \mathfrak{D},$$

where

$$\mathfrak{d}_{31} = \mathfrak{c}_1, \qquad \mathfrak{d}_{32} = \mathfrak{c}_2, \qquad \mathfrak{d}_{33} = \mathfrak{c}_3,$$
$$\mathfrak{d}_{21} = \mathfrak{b}_3\mathfrak{c}_1, \qquad \mathfrak{d}_{22} = \mathfrak{b}_3\mathfrak{c}_2 + \mathfrak{b}_2, \qquad \mathfrak{d}_{23} = \mathfrak{b}_3\mathfrak{c}_3 + \mathfrak{b}_1,$$
$$\mathfrak{d}_{11} = \mathfrak{a}_2\mathfrak{d}_{21} + \mathfrak{a}_3\mathfrak{c}_1, \qquad \mathfrak{d}_{12} = \mathfrak{a}_1 + \mathfrak{a}_2\mathfrak{d}_{22} + \mathfrak{a}_3\mathfrak{c}_2, \qquad \mathfrak{d}_{13} = \mathfrak{a}_2\mathfrak{d}_{23} + \mathfrak{a}_3\mathfrak{c}_3.$$

Now provided that the determinant $|\mathfrak{D}| \neq 0$, we can solve these nine equations for the $\mathfrak{a}_1, \mathfrak{a}_2, \ldots, \mathfrak{c}_3$. This proves the theorem.

The proof in the m-ary case is similar.

It follows from this theorem that any function which remains invariant under every transformation of the above three types is invariant under every transformation of the group. Also if the group is Γ then the $\mathfrak{k}$ in Type II must be an element of the $GF[p^n]$. Thus a formal covariant is invariant under Type II if, and only if, $\mathfrak{k}$ is an element of the $GF[p^n]$, unless it be also a congruent covariant.

§ 8. Weight and Isobarism.

Following the method of Elliott* we define the term weight. Suppose that our ground form is an m-ary l-ic in the m variables $x_1, \ldots, x_m$, and let the coefficients of

$x_1^l,\ x_2^l,\ \ldots,\ x_{m-1}^l,\ x_1^{l-1}x_2,\ \ldots$	have the suffix	0,
$x_1^{l-1}x_m,\ x_2^{l-1}x_m,\ \ldots,\ x_1^{l-2}x_2x_m,\ \ldots$	,,	1,
$x_1^{l-2}x_m^2,\ \ldots,\ x_1^{l-3}x_2x_m^2,\ \ldots$	,,	2,
....	....	
$x_1x_m^{l-1},\ x_2x_m^{l-1},\ \ldots,\ x_{m-1}x_m^{l-1},\ \ldots$	,,	$l-1$,

x_m^l, have the suffix l,

that is, the suffix of any coefficient is equal to the power of x_m which it multiplies in the form. In addition we say that the suffix of x_m is 0, while the suffix of x_i is 1 for $i \neq m$. The WEIGHT of any term is defined as being the sum of the suffixes of its various factors. Thus the weight of each term in the ground form is l. If a polynomial in the coefficients a and the variables x be such that each term is of the same weight w, then the polynomial is said to be ISOBARIC. This is a definition of "isobaric" according to Elliott, but we note that to state quite clearly what we mean we must use the phase "isobaric with respect to x_m." We shall therefore make the following definition: If a polynomial is isobaric with respect to all variables, it is said to be COMPLETELY ISOBARIC.

For example suppose that $f = a_1x_1 + a_2x_2 + a_3x_3$ be the ground form: then $g = a_1x_1 + a_3x_2 + a_2x_3$ is isobaric with respect to x_1 but is not isobaric with respect to x_2 or x_3 and is therefore not completely isobaric. By considering transformations of Type III we see that if a concomitant is isobaric with respect to any variable, it is completely isobaric. Thus g could not be a covariant of f.

§ 9. Congruent Concomitants.

By definition it is clear that congruent concomitants only differ from the algebraic concomitants in that it is permitted in the former case to make the modular reduction $p \equiv 0$. Let us take as our ground form

$$\phi = \alpha_x\beta_x, \ldots, \delta_x, \text{ where } \alpha_x = \alpha_1x_1 + \ldots + \alpha_mx_m,$$

* *Algebra of Quantics* (first edition), p. 38.

and where there are r symbols $\alpha, \beta, \ldots, \delta$. Now if C be a congruent covariant of ϕ where C is a function of variables x and the non-symbolical coefficients a, then we can write

$$C(\bar{a}, \bar{x}) \equiv M(\mathfrak{a})\, C(a, x),$$

where M is a function of the coefficients of the transformation only. As in the algebraic case* we can shew that C is homogeneous in the variables x, or else is the sum of covariants which are homogeneous in the variables x. We can assume then without loss of generality that C is homogeneous in the variables x.

Now considering transformations whose matrices are of the type

$$\begin{bmatrix} \mathfrak{k} & \cdot & & \cdot \\ \cdot & \mathfrak{k} & & \cdot \\ & & \ddots & \\ \cdot & \cdot & & \mathfrak{k} \end{bmatrix},$$

where $\mathfrak{k}$ is a non-zero scalar, we have easily that C must be homogeneous in the non-symbolical coefficients a. By considering transformations of the type

$$x_i = \bar{x}_j,$$
$$x_j = \bar{x}_i,$$
$$x_k = \bar{x}_k \qquad (k \neq i,\ k \neq j),$$

we see that C must be symmetrical with respect to the suffixes both of the x's and of the symbols $\alpha, \beta, \ldots, \delta$.

Let us represent then our covariant $C(a, x)$ symbolically. A single term of this representation can be written as follows:

$$c_i = x_1^{s_1} x_2^{s_2} \ldots x_m^{s_m} \alpha_1^{h_1} \ldots \alpha_m^{h_m} \beta_1^{k_1} \ldots \beta_m^{k_m} \ldots \delta_1^{l_1} \ldots \delta_m^{l_m},$$

where $\qquad S = s_1 + s_2 + \ldots + s_m$

and $\qquad H = h_1 + h_2 + \ldots + h_m = k_1 + \ldots + k_m = \ldots = l_1 \ldots l_m$

are constants for every term c_i of C.

Now if $\qquad x_1 = \mathfrak{k}\bar{x}_1,$

$\qquad x_i = \bar{x}_i \qquad (i \neq 1),$

then $\qquad \bar{x}_1 = \mathfrak{k}^{-1} x_1 \qquad$ and $\bar{\alpha}_1 = \mathfrak{k}\alpha_1,$

$\qquad \bar{x}_i = x_i \qquad (i \neq 1), \qquad \bar{\alpha}_i = \alpha_i \qquad (i \neq 1),$

so that $\qquad \bar{c}_i = \mathfrak{k}^T c_i$, where $T = h_1 + k_1 + \ldots + l_1 - s_1$.

Now $\mathfrak{k}^T$ must be the same for every term c_i, so that since $\mathfrak{k}$ is an arbitrary non-zero scalar we have that T is a constant and therefore

* Elliott, *loc. cit.* p. 40.

$T+S$ is a constant, but $T+S$ is the weight with respect to x_1, so that C is isobaric with respect to x_1 and therefore is completely isobaric. This gives us the theorem: *A congruent concomitant is completely isobaric.* Conversely if a formal concomitant is isobaric, then $T+S$ is a constant and so is T, and so the concomitant is invariant for all values of $\mathfrak{k}$ and is therefore a congruent concomitant. Miss Hazlett* proved the converse for the binary case. We shall also extend to the m-ary case her proof of the following theorem†: *If a congruent concomitant has factors, these also are isobaric.* For, let C have factors C_1 and C_2 which are not both isobaric, then the term which has the greatest/least weight in the product $C_1 \times C_2$ is the product of the terms which have the greatest/least weight in C_1 and C_2 respectively. The theorem follows at once.

By considering transformations whose matrices are

$$\begin{bmatrix} \mathfrak{k} & . & & . \\ . & \mathfrak{k} & & . \\ & & \ddots & \\ . & . & & \mathfrak{k} \end{bmatrix} \text{ and } \begin{bmatrix} \mathfrak{h} & . & & . \\ . & 1 & & . \\ & & \ddots & \\ . & . & & 1 \end{bmatrix},$$

we find as in the algebraic case that the following relations hold for congruent concomitants:

$$rH - S = mK,$$
$$rH + (m-1)S = mW_i \quad \text{..................(9·1)},$$

where the index is K, the order S, the degree H and the weight is W_i with respect to any variable x_i and where r is the degree of the ground form.

§ 10. Relation between Congruent and Algebraic Covariants.

Let $f = ax_1^2 + 2bx_1x_2 + cx_2^2$. It is well known that $b^2 - ac$ is an algebraic invariant of this form. It is therefore also a congruent invariant. Also $(b^2-ac)^3$ must be both an algebraic and a congruent invariant, but if $p = 3$, then

$$(b^2-ac)^3 \equiv b^6 - a^3c^3$$

and so $b^6 - a^3c^3$ must be a congruent invariant. It is not, however, an algebraic invariant. Thus it would seem that for one algebraic invariant $(b^2-ac)^3$ we have four congruent invariants, viz.

$$(b^2-ac)^3, \quad b^6 - 3b^4ac - a^3c^3, \quad b^6 + 3b^2a^2c^2 - a^3c^3, \quad b^6 - a^3c^3;$$

these are all, however, congruent to each other, and therefore represent

* *Trans. Amer. Math. Soc.* vol. 24, p. 296 (1922).
† *Ibid.* p. 297.

the same congruent invariant. It will be more convenient therefore if we neglect any term in a congruent covariant having the factor p. If we proceed in this way we infer that one, and only one, congruent covariant is obtained from any algebraic covariant, and we can therefore represent such a covariant symbolically since every algebraic covariant can be represented symbolically. We now put the important question. Can every congruent covariant be represented symbolically? In other words, does there correspond to every congruent covariant an algebraic covariant? An answer to this question for the general case has not yet been given, but clearly it forms the keystone of the symbolical theory of modular covariants. A symbolical theory is obviously not of much use if covariants exist which cannot be represented symbolically. It is extremely likely that every congruent covariant can be represented symbolically, but in the absence of proof we shall have to divide congruent covariants into two sorts, symbolical congruent covariants and non-symbolical congruent covariants, i.e. those which cannot be represented symbolically. If it be proved later that this second kind does not exist, then our discussion of the first kind will be applicable to all congruent covariants.

Let C be a congruent covariant: then will $\overline{C} \equiv MC$, where M is a function of the coefficients. Let C be a sum of terms P_i, thus $C = \sum_{i=1}^{a} P_i$ and similarly $\overline{C} = \sum_{i=1}^{a} \overline{P}_i = \sum_{i=1}^{b} M_i P_i$, say, so that, for all values of the $\mathfrak{a}$'s,

$$\sum_{i=1}^{b} M_i P_i \equiv \sum_{i=1}^{a} M P_i,$$

then must $$M_i \equiv 0 \quad \text{if } i > a$$

and $$M_i \equiv M \quad \text{if } i \leqslant a.$$

We prove that M is congruent to a power of $|\mathfrak{A}|$ by the same argument as that used in the algebraic case. The proof of the binary algebraic case given by Grace and Young* can be extended to the m-ary case. Thus each M_i $(i = 1, \ldots, b)$ is a power of the determinant of the transformation or else zero and is therefore isobaric since Fermat's Theorem cannot be applied here. Miss Hazlett† proved this for the binary case.

Miss Hazlett and R. Weitzenböck have proved in special cases that every congruent covariant can be represented symbolically.

* *Algebra of Invariants*, p. 22.

† *Trans. Amer. Math. Soc.* vol. 24, p. 297 (1922).

R. Weitzenböck's proof is as follows:

Let C be a congruent covariant, then as before

$$C(\bar{a}, \bar{x}) \equiv M(\mathfrak{a})\, C(a, x) \quad \text{and} \quad M(\mathfrak{a}) \equiv |\mathfrak{A}|^s,$$

or
$$C(\bar{a}, \bar{x}) = |\mathfrak{A}|^s\, C(a, x) + pD(a, x, \mathfrak{a}).$$

Now operate s times with the Cayley operator Ω. Then $\Omega^s |\mathfrak{A}|^s = c_s$ a constant so that $\Omega^s C(\bar{a}, \bar{x})$ is independent of $\mathfrak{a}$ and is therefore* an algebraic covariant K, say. Now $\Omega^s D(a, x, \mathfrak{a})$ must also be independent of $\mathfrak{a}$ and we shall call this expression E. Then

$$K = c_s C(a, x) + pE \quad \text{or} \quad K \equiv c_s C(a, x).$$

Now if c_s is not divisible by p, then $C(a, x) \equiv \dfrac{K}{c_s}$, which is an algebraic covariant and is therefore representable symbolically; so that $C(a, x)$ is also representable symbolically. Unfortunately c_s is very often divisible by p, in which case the proof does not hold. In fact†

$$c_s = \frac{\underline{|m+s-1}\ \ \underline{|m+s-2} \dots \underline{|m}}{\underline{|s-1}\ \ \underline{|s-2} \dots \underline{|0}}$$

and is therefore divisible by p unless $m + s - 1 < p$, i.e. unless

$$s < p - m + 1.$$

Miss Hazlett‡ proved as in the theory of algebraic seminvariants that if I be a congruent seminvariant of a system S of binary forms, which is of degree g and weight w, then I is congruent, modulo p, to a product of a power of a_0 and a symmetric polynomial P in symbolic ratios $\dfrac{a_2}{a_1}, \dfrac{\beta_2}{\beta_1}, \dots$, which is homogeneous in these ratios. Moreover P is expressible as a polynomial in the differences of these ratios. From this she proved§ that every congruent invariant of a system of binary forms is congruent to an algebraic invariant. The symbolical representation of such invariants follows immediately.

Since symbolical congruent covariants are congruent to algebraic covariants, they can be represented by the corresponding symbolical expressions. The finiteness of symbolical congruent covariants follows at once from the algebraic case. It is seen that such symbolical congruent covariants are very similar to the algebraic case, and we obtain full systems of the one from the full systems of the other.

* Weitzenböck, *Invariantentheorie*, p. 147.

† Weitzenböck, *loc. cit.* p. 16.

‡ *Trans. Amer. Math. Soc.* vol. 24, p. 298 (1922).

§ *Trans. Amer. Math. Soc.* vol. 30, p. 855 (1928).

§ 11. Formal Covariants.

Not every formal covariant is isobaric; and so while every congruent covariant is also a formal covariant the converse is not true. As in § 9, $\mathfrak{k}^T$ must take the same value for every term c_i of $C(a, x)$, but according to definition $\mathfrak{k}$ is no longer an arbitrary non-zero scalar but is now an element of $GF[p^n]$ and we may therefore reduce $\mathfrak{k}^T$ by Fermat's Theorem. Thus it is no longer necessary for T to be equal to a constant but T must be congruent to a constant modulo q for every term c_i. It follows that the weights with respect to all variables must be congruent modulo q to each other. Thus the weights of the different terms of a formal covariant must be congruent modulo q. In this case we say that a formal covariant which is not isobaric is PSEUDO-ISOBARIC. The universal covariants which we shall now discuss are all pseudo-isobaric, and since they are independent of the coefficients of the form they are also residual covariants, without any reduction being performed.

The equations (9·1) must hold also for formal concomitants if we replace the equal sign by a congruence modulo q:

thus
$$\left.\begin{aligned} rH - S &\equiv mK \pmod{q}, \\ rH + (m-1)S &\equiv mW_i \pmod{q}. \end{aligned}\right\} \quad \text{.........(11·1).}$$

§ 12. Universal Covariants*.

Let $x_1, \ldots, x_m$ be a set of variables which undergo the following transformation:

$$x_i = \sum_{j=1}^{m} \mathfrak{a}_{ij} \bar{x}_j \qquad (i = 1, \ldots, m),$$

where each $\mathfrak{a}_{ij}$ belongs to $GF[p^n]$ and where

$$|\mathfrak{a}_{ij}| \neq 0;$$

then
$$x_i^{p^n} = \sum_{j=1}^{m} (\mathfrak{a}_{ij}\bar{x}_j)^{p^n} + p^n (\ldots)$$

or
$$x_i^{p^n} \equiv \sum_{j=1}^{m} \mathfrak{a}_{ij}\bar{x}_j^{p^n} \text{ and thus } x_i^{p^{nt}} \equiv \sum_{j=1}^{m} \mathfrak{a}_{ij}\bar{x}_j^{p^{nt}}.$$

We have therefore the important result that the set $x_1^{p^{nt}}, \ldots, x_m^{p^{nt}}$ is modularly cogredient with the set $x_1, \ldots, x_m$, where t is any positive integer. Now let us write

$$[e_1, \ldots, e_m] \equiv \begin{vmatrix} x_1^{p^{e_1 n}} & \ldots & x_m^{p^{e_1 n}} \\ \vdots & & \vdots \\ x_1^{p^{e_m n}} & \ldots & x_m^{p^{e_m n}} \end{vmatrix}.$$

* Dickson, *Trans. Amer. Math. Soc.* vol. 12, p. 75 (1911).

As in the case of algebraic invariants this must be an invariant since all the rows are cogredient. $[e_1, \ldots, e_m]$ is an example of a universal covariant; it is frequently referred to also as an invariant of the group since it is independent of any ground form. There is no expression in the theory of algebraic invariants corresponding to this, if there are less than m sets of cogredient variables.

In particular we shall write

$$[m, m-1, \ldots, s+1, s-1, \ldots, 1, 0] = L_m^{(s)}, \quad L_m^{(m)} = L_m, \quad \frac{L_m^{(s)}}{L_m} = Q_{m,s}.$$

We have seen in § 4 that all points with coordinates belonging to $GF[p^n]$ are conjugate under Γ_1. Thus if a covariant vanishes when the x's take the values of one of these points, it must vanish for all; therefore if a covariant contains one factor of

$$E_m = \prod_{k=1}^{m} \prod_{c_i/p^n} (x_k + c_{k+1} x_{k+1} + \ldots + c_m x_m),$$

where c_i/p^n denotes that each c_i in the product takes all the p^n possible values, it will contain every factor and therefore E_m itself. Now L_m has the factor x_m and therefore the factor E_m, and comparing coefficients we have that

$$L_m \,|||\, E_m.$$

Similarly every $L_m^{(s)}$ has the factor x_m and is therefore either zero or has L_m as a factor: hence $Q_{m,s}$ is rational and is an absolute covariant since the index of every $L_m^{(i)}$ is 1. Also L_{m-1} divides L_m since E_{m-1} is obviously a factor of E_m.

Consider the following determinant which vanishes identically.

$$\left|\begin{array}{cccccc}
x_1^{p^{nm}} & \ldots & x_m^{p^{nm}} & x_1^{p^{nm}} & \ldots & x_{m-1}^{p^{nm}} \\
x_1^{p^{n(m-1)}} & \ldots & x_m^{p^{n(m-1)}} & x_1^{p^{n(m-1)}} & \ldots & x_{m-1}^{p^{n(m-1)}} \\
\cdots & \cdots & \cdots & \cdots & \cdots & \cdots \\
x_1^{p^n} & \ldots & x_m^{p^n} & x_1^{p^n} & \ldots & x_{m-1}^{p^n} \\
x_1 & \ldots & x_m & x_1 & \ldots & x_{m-1} \\
0 & \ldots & 0 & x_1^{p^{n(m-1)}} & \ldots & x_{m-1}^{p^{n(m-1)}} \\
\cdots & \cdots & \cdots & \cdots & \cdots & \cdots \\
\cdots & \cdots & \cdots & x_1^{p^{n(s+1)}} & \ldots & x_{m-1}^{p^{n(s+1)}} \\
\cdots & \cdots & \cdots & x_1^{p^{n(s-1)}} & \ldots & x_{m-1}^{p^{n(s-1)}} \\
\cdots & \cdots & \cdots & \cdots & \cdots & \cdots \\
\cdots & \cdots & \cdots & \cdots & \cdots & \cdots \\
0 & \ldots & 0 & x_1^{p^n} & \ldots & x_{m-1}^{p^n}
\end{array}\right|
\begin{array}{l} \left.\vphantom{\begin{array}{c}a\\a\\a\\a\\a\end{array}}\right\} m+1 \text{ rows} \\ \left.\vphantom{\begin{array}{c}a\\a\\a\\a\\a\\a\\a\end{array}}\right\} m-2 \text{ rows} \end{array}$$

$\underbrace{\qquad\quad}_{m \text{ cols.}} \qquad \underbrace{\qquad\quad}_{m-1 \text{ cols.}}$

By Laplace's development we get

$$\begin{aligned}&(-1)^{m-2}[m, \ldots, 1][m-1, \ldots, s+1, s-1, \ldots, 1, 0]\\ &+(-1)^{m-1}[m, \ldots, s+1, s-1, \ldots, 0][m-1, \ldots, 1]\\ &+(-1)^{m}\ \ [m-1, \ldots, 0][m, \ldots, s+1, s-1, \ldots, 1]=0.\end{aligned}$$

If therefore $1 < s < m-1$ we have

$$0 = L_m^{p^n} Q_{m-1,s} L_{m-1} - Q_{m,s} L_m L_{m-1}^{p^n} + L_m (Q_{m-1,s-1} L_{m-1})^{p^n},$$

and so

$$Q_{m,s} = Q_{m-1,s}\left(\frac{L_m}{L_{m-1}}\right)^{p^n-1} + Q_{m-1,s-1}^{p^n} \quad \ldots\ldots\ldots(12\cdot 1).$$

For the case where $s=1$ we find that the exponents in the last row of the determinants are equal to p^{2n}. If $s = m-1$ the exponents of the $(m+2)$th row are $p^{n(m-2)}$ and we have

$$L_m^{p^n} Q_{m-1,1} L_{m-1} - Q_{m,1} L_m L_{m-1}^{p^n} + L_m L_{m-1}^{p^{2n}} = 0,$$

so that

$$Q_{m,1} = Q_{m-1,1}\left(\frac{L_m}{L_{m-1}}\right)^{p^n-1} + L_{m-1}^{p^{2n}-p^n} \quad \ldots\ldots(12\cdot 2),$$

and

$$L_m^{p^n} L_{m-1} - Q_{m,m-1} L_m L_{m-1}^{p^n} + L_m (Q_{m-1,m-2} L_{m-1})^{p^n} = 0,$$

so that

$$Q_{m,m-1} = \left(\frac{L_m}{L_{m-1}}\right)^{p^n-1} + Q_{m-1,m-2}^{p^n} \quad \ldots\ldots\ldots(12\cdot 3).$$

Expanding L_m by the last column we have

$$L_m = x_m L_{m-1}^{p^n} + L_{m-1}\sum_{s=1}^{m-2}(-1)^s x_m^{p^{ns}} Q_{m-1,s} + (-1)^{m-1} x_m^{p^{n(m-1)}} L_{m-1} \quad \ldots\ldots(12\cdot 4).$$

Thus L_m is divisible by L_{m-1} and each expression (12·1), (12·2), (12·3) can be given in an integral form.

The degree of $Q_{m,s}$ is $p^{nm} - p^{ns}$ and that of L_m is

$$p^{n(m-1)} + \ldots + p^n + 1 \quad \ldots\ldots\ldots\ldots\ldots\ldots\ldots(12\cdot 5).$$

§ 13. Dickson's Theorem.

Dickson* proved the finiteness theorem for universal covariants and gave a full system. The proof of this theorem is of course included in E. Noether's Theorem (*v.* § 43), but we require Dickson's proof however in order to obtain the full system. The theorem is stated as follows:

The functions L_μ, $Q_{\mu,1}$, ..., $Q_{\mu,\mu-1}$ are independent and form a full system of the invariants of Γ^μ.

* *Trans. Amer. Math. Soc.* vol. 12, pp. 75–98 (1911).

The theorem is proved by induction. We assume that it is true for $\mu \leqq m$ where $m \geqq 2$, and prove that it is true for $\mu = m+1$. Let I be any homogeneous integral universal covariant of Γ^{m+1}. The coefficients of the various powers in x_{m+1} in I must be universal covariants of Γ^m, and hence by hypothesis are integral functions of $L_m, \ldots, Q_{m, m-1}$. We suppose then that

$$I = I_0 + x_{m+1} I_1 + \ldots + x_{m+1}^r I_r,$$

where each I_i is an integral function of $L_m, \ldots, Q_{m, m-1}$. Then I_0 is an aggregate of terms

$$t' = c' L_m^{a'} Q_{m,1}^{b_1'} \ldots Q_{m, m-1}^{b'_{m-1}}.$$

Consider those terms in which a' is a minimum a, the sub-set where b' is a minimum b, and so on, so that finally we obtain the unique term

$$t = c L_m^a Q_{m,1}^{b_1} \ldots Q_{m, m-1}^{b_{m-1}}.$$

Expanding by (12·1), ..., (12·5), we see that the terms of minimum degree in x_m are included in

$$x_m^a t_1 \quad \text{where } t_1 = c L_{m-1}^{a_1} \prod_{s=2}^{m-1} Q_{m-1, s-1}^{p^n b_s} \qquad (a_1 = ap^n + b_1 d).$$

Similarly the terms of the minimum degree in x_{m-1} are included in

$$x_{m-1}^{a_1} t_2 \quad \text{where } t_2 = c L_{m-2}^{a_2} \prod_{s=3}^{m-1} Q_{m-2, s-2}^{p^{2n} b_s} \qquad (a_2 = a_1 p^n + p^n b_2 d),$$

and so on, so that finally t contains a term

$$\tau = c x_m^a x_{m-1}^{a_1} \ldots x_1^{a_{m-1}} \qquad (a_i = a_{i-1} p^n + p^{n(i-1)} b_i d):$$

and this term τ will only occur once in t and not in any other product except t. Now I has the isolated term τ and therefore also

$$x_{m+1}^{a_1} \tau_1 \quad \text{where } \tau_1 = c x_{m-1}^a x_{m-2}^{a_2} \ldots x_1^{a_{m-1}}.$$

Therefore τ_1 is a term of a universal covariant of Γ^m. Similarly

$$\tau_2 = c x_{m-2}^a x_{m-3}^{a_3} \ldots x_1^{a_{m-1}}$$

is a term of a universal covariant of Γ^{m-1} and $c x_1{}^a$ is a term of a universal covariant of Γ^2, therefore $c x_1{}^a$ is a term of $k Q_{21}{}^a$ whence $a = \alpha d$.

By (12·5) the degree of $Q_{m,s}$ is a multiple of p^n, and since a is a multiple of p^n, therefore the degree of t is a multiple of p^n. This is therefore true of t', and since the degree of L_m is prime to p therefore a' is a multiple of p^n; it is also a multiple of q, for apply a transformation of determinant ρ to I_0, then the Q's being absolutely invariant we have $\rho^{a'} \parallel \rho^a \bmod p$, hence $a' \equiv a \bmod q$; but a is divisible by d and therefore by q, so that a' is also divisible by d. Hence in every term t' of I_0, a' is a multiple of d. From (12·1)—(12·4) we have

$$Q_{m,1} = L^{d}_{m-1} + x^{q}_{m} L^{q^2}_{m-1} Q_{m-1,1} + \ldots$$

and
$$Q_{m,s} = Q^{p^n}_{m-1,s-1} + x^{q}_{m} L^{q^2}_{m-1} Q_{m-1,s} + \ldots \qquad (s > 1)$$

(the factor $Q_{m-1,s}$ being suppressed if $s = m-1$).

In these series the exponents of x_m differ by multiples of q. Let $b_s = p^{\beta_s} B_s$, where B_s is prime to p. To obtain the p^{β}th power of a sum in a field having modulus p we have only to multiply every exponent by p^{β}. Hence

$$L^{a}_{m} = (L^{p^n}_{m})^{aq} = x^{a}_{m} L^{ap^n}_{m-1} - aq x^{a+d}_{m} L^{ap^n-d}_{m-1} Q^{p^n}_{m-1,1} + \ldots \qquad \ldots\ldots(13·1),$$

$$Q^{b_1}_{m,1} = L^{db_1}_{m-1} + B_1 x^{qp^{\beta_1}}_{m} L^{e}_{m-1} Q^{p^{\beta_1}}_{m-1,1} + \ldots, \qquad \text{where } e = db_1 - qp^{\beta_1} \ \ldots\ldots(13·2),$$

$$Q^{b_s}_{m,s} = Q^{p^n b_s}_{m-1,s-1} + B_s x^{qp^{\beta_s}}_{m} L^{q^2 p^{\beta_s}}_{m-1} Q^{e_s}_{m-1,s-1} Q^{p^{\beta_s}}_{m-1,s} + \ldots, \qquad \text{where } e_s = b_s p^n - p^{n+\beta_s} \ \ldots\ldots(13·3),$$

where in the last series $s > 1$ and the factor $Q_{m-1,s}$ is to be suppressed if $s = m-1$. Hence t/c contains the terms

$$T_1 = B_1 x^{a+qp^{\beta_1}}_{m} L^{e+ap^n}_{m-1} Q^{p^{\beta_1}}_{m-1,1} \prod_{s=2}^{m-1} Q^{p^n b_s}_{m-1,s-1},$$

$$T_\sigma = B_\sigma x^{a+qp^{\beta_\sigma}}_{m} L^{h_\sigma}_{m-1} Q^{e_\sigma}_{m-1,\sigma-1} Q^{p^{\beta_\sigma}}_{m-1,\sigma} \prod_{s} Q^{p^n b_s}_{m-1,s-1} \qquad (h_\sigma = ap^n + db_1 + q^2 p^{\beta_\sigma}),$$

where $\sigma > 1$, and $Q_{m-1,\sigma}$ is to be suppressed if $\sigma = m-1$, and where in the final product s has the values $2, \ldots, \sigma-1, \sigma+1, \ldots, m-1$.

First let $\beta_1 < n$, then $qp^{\beta_1} < d$. The product t contains but one term with the same set of exponents as T_1. For if we employ a term of (13·2) after the second, the exponent of x_m exceeds that in T_1. If we employ the second term in (13·2) we must use the first terms in (13·1) and (13·3) and hence get T_1 itself. If we employ the first term of (13·2) we must use the first term of (13·1), and obtain $ap^n + db_1$ as the exponent of L_{m-1} in the product of the two, which is greater than $e + ap^n$ which is impossible. Suppose that T_1 is a term of a product t' distinct from t. If $a' > a$, then $a' \geqslant a + d$, since a' and a are multiples of d. Hence the minimum exponent a' of x_m in t' would exceed the exponent of x_m in T_1. It follows that $a' = a$, further by (11·5) and by the homogeneity of the universal covariant,

$$\sum_{s=1}^{m-1} (b_s' - b_s)(p^{nm} - p^{ns}) = 0 \quad \text{.............. (13·4).}$$

Hence $(b_1' - b_1)p^n$ is a multiple of p^{2n}, so that $b_1' \equiv b_1 \bmod p^n$ when $a' = a$. Thus in $b_1' = p^{\beta_1'} B_1'$ we have, since β_1 and β_1' are both less than n,

$$\beta_1' = \beta_1 \quad \text{.......................... (13·5).}$$

T_1 cannot therefore occur in terms of t' other than

$$x_m^a L_{m-1}^{ap^n} (L_{m-1}^{db_1'} + B_1' x_m^{qp^{\beta_1}} L_{m-1}^{e'} Q_{m-1,1}^{p^{\beta_1}}) \prod_{s=2}^{m-1} Q_{m,s}^{b_s'} \quad \text{...... (13·6).}$$

If we employ the second term in the parenthesis we must take the term of each $Q_{m,s}$ free of x_m. Then $b_1' = b_1$, from the exponents of L_{m-1}, and $b_s' = b_s$ $(s = 2, \ldots, m-1)$, from the exponents of $Q_{m-1, s-1}$. But this is impossible, for $t' \neq t$. If we employ the first term in the parenthesis in (13·6) we obtain as the exponent of L_{m-1} in the product of the first two factors $ap^n + db_1' > e + ap^n$, since $b_1' \geqq b_1$ when $a' = a$. Hence our assumption is false. We have now shown that T_1 occurs as an isolated term of the invariant. But, the exponent of x_m is not a multiple of d, and the coefficient B_1 is not zero. Hence the case $\beta_1 < n$ is excluded and therefore b_1 is a multiple of p^n.

Of the numbers $b_2, \ldots, b_{m-1}$ not multiples of p^n let b_σ be the one with the smallest subscript. A term of t with the same set of exponents as T_σ can be obtained only by taking the first term of (13·1), (13·2), (13·3) for $s < \sigma$, for otherwise the exponent of x_m is divisible by d. If we use the second term of (13·3) for $s = \sigma$, we must use the first term of (13·3) for $s > \sigma$ and then obtain T_σ. If we use the first term of (13·3) for $s = \sigma$ the exponent of $Q_{m-1, \sigma-1}$ in the product is $p^n b_\sigma$, which exceeds its exponent e_σ in T_σ. Next if T_σ occurs in t' distinct from t, then $a' = a$. From (13·5) b_1' is a multiple of p^n. Analogous to (13·2)

$$Q_{m,1}^{b_1'} = L_{m-1}^{db_1'} + x_m^d K \quad \text{................. (13·7).}$$

Hence we take the first terms of (13·1) and (13·7). In the product of these two, the exponent of L_{m-1} is $ap^n + db_1' > h_\sigma$ if $b_1' \geqslant b_1 + p^n$. It follows that $b_1' = b_1$. If $\sigma > 2$, b_2 is by hypothesis a multiple of p^n and so is b_2' by (13·4), and we must take therefore the first term $Q_{m-1,s-1}^{p^n b_2'}$ of $Q_{m,2}^{b_2'}$. Since $Q_{m-1,1}$ does not occur in the expansion of $Q_{m,s}$ for $s > 2$ but occurs in T_σ with the exponent $p^n b_2$, we conclude that $b_2' = b_2$. In this manner we can show that we must take the first term of $Q_{m,s}^{b_s'}$ $(s < \sigma)$ and that $b_s' = b_s$

$(s=2, \ldots, \sigma-1)$. Then by (13·4) $b_\sigma' \equiv b_\sigma \bmod p^n$ whence $\beta_\sigma' = \beta_\sigma$. If we employ the second term in $Q^{b_\sigma'}_{m,\sigma}$, we must use the first term in $Q^{b_s'}_{m,s}$ $(s>\sigma)$ and we obtain T_σ if, and only if, $b_s' = b_s$ $(s=\sigma, \ldots, m-1)$, as shown by comparing the exponents of $Q_{m-1,s}$ $(s \geqslant \sigma)$, but in this case $t' \equiv t$. If we employ the first term in $Q^{b_\sigma'}_{m,\sigma}$, the total exponent of $Q_{m-1,\sigma-1}$ in t' is $p^n b_\sigma'$ which exceeds its exponent e_σ in T_σ since $b_\sigma' \geqslant b_\sigma$ in view of our definition of t.

We have now shown that T_σ occurs as an isolated term of the invariant. But the exponent of x_m is not a multiple of d since b_σ is not a multiple of p^n, and the coefficient B_σ is not zero in the field. Hence our assumption concerning b_σ is false, so that $b_1, \ldots, b_{m-1}$ are all multiples of p^n. Put $b_s = p^n c_s$, then

$$I' = I - cQ^{\sigma}_{m+1,1} \prod_{s=1}^{m-1} Q^{c_s}_{m+1,s+1}$$

is an invariant of Γ^{m+1} in which I_0' lacks t. We can proceed in this manner until we have an invariant of Γ^{m+1} which has no terms free of x_{m+1}, it therefore has L_{m+1} as a factor. Proceed as before and we infer finally that any integral universal covariant of Γ^{m+1} is an integral function of L_{m+1}, $Q_{m+1,i}$ $(i=1, \ldots, m)$.

As a basis of the induction Dickson* has proved that any integral invariant with coefficients in the $GF[p^n]$ of the group Γ^2 is an integral function of Q_{21} and L_2 with coefficients in the $GF[p^n]$.

We have still to show that these universal covariants are independent. For if they are not, then there will be a relation

$$L_{m+1} A(L_{m+1}, Q_{m+1,1}, \ldots, Q_{m+1,m}) + B(Q_{m+1,1}, \ldots, Q_{m+1,m}) = 0 \qquad \ldots\ldots(13·8)$$

Putting $x_{m+1}=0$ and substituting for the Q's in B we have

$$B(L^d_m, Q^{p^n}_{m,1}, \ldots, Q^{p^n}_{m,m-1}) = 0,$$

but L_m and the $Q_{m,s}$ are independent by hypothesis and so $B \equiv 0$, so that the relation (13·8) has the factor L_{m+1}. Removing this factor and repeating the above process we prove every successive $B=0$ and the covariants $L_{m+1}, Q_{m+1,1}, \ldots, Q_{m+1,m}$ independent.

As a basis of the induction proof used above we notice that there is no relation $AL_2 + CQ^e_{2,1} = 0$ between the universal covariants of Γ^2, for by putting $x_2 = 0$ we have $C=0$ and finally $A=0$.

This concludes the proof of the important theorem.

It seems almost certain that a similar theorem would hold for the

* *Trans. Amer. Math. Soc.* vol. 12, p. 4 (1911).

case of several cogredient variables x, y, z, ... and that a full system of universal covariants in such a case would contain determinants and quotients of determinants in which only the following rows would appear:

$$\begin{array}{lll} x_1 \ \dots \ x_m, & y_1 \ \dots \ y_m, & \\ x_1^{p^n} \ \dots \ x_m^{p^n}, & y_1^{p^n} \ \dots \ y_m^{p^n}, & \text{etc.} \\ x_1^{p^{nt}} \ \dots \ x_m^{p^{nt}}, & y_1^{p^{nt}} \ \dots \ y_m^{p^{nt}}, & \end{array}$$

The proof of such a theorem for the perfectly general case by the above method would be cumbersome in the extreme. W. C. Krathwohl* has proved a theorem of this sort for the binary case where there were two sets of variables. The full system which he gave is, with slight changes of notation, as follows:

$$L_x = \begin{vmatrix} x_1^p & x_2^p \\ x_1 & x_2 \end{vmatrix}, \qquad L_y = \begin{vmatrix} y_1^p & y_2^p \\ y_1 & y_2 \end{vmatrix},$$

$$M = \begin{vmatrix} x_1 & x_2 \\ y_1 & y_2 \end{vmatrix}, \qquad M_1 = \begin{vmatrix} x_1^p & x_2^p \\ y_1 & y_2 \end{vmatrix}, \qquad M_2 = \begin{vmatrix} x_1 & x_2 \\ y_1^p & y_2^p \end{vmatrix},$$

$$Q_x = \frac{\begin{vmatrix} x_1^{p^2} & x_2^{p^2} \\ x_1 & x_2 \end{vmatrix}}{L_x}, \qquad Q_y = \frac{\begin{vmatrix} y_1^{p^2} & y_2^{p^2} \\ y_1 & y_2 \end{vmatrix}}{L_y},$$

$$N_s = \frac{M_2^{s+1} L_x^{p-s-1} + (-)^s M_1^{p-s} L_y^s}{M^p} \qquad (1 \leqslant s \leqslant p-2).$$

§ 14. Formal Invariants of the Linear Form.

Although the theorem of the previous paragraph is strictly concerned with universal covariants only, it at once furnishes us with a method for obtaining a large number of formal invariants and covariants. If $a_1, \dots, a_m$ be any set of quantities which is contragredient to $x_1, \dots, x_m$, then so also is the set $a_1^{p^n}, \dots, a_m^{p^n}$ contragredient; and we at once obtain the following pure invariants of the linear ground form $a_1 x_1 + \dots + a_m x_m$,

$$L_{(a)m}, \ Q_{(a)m,s} \qquad (s = 1, \dots, p-1) \ \dots\dots\dots(14{\cdot}1),$$

where $L_{(a)m}$ is obtained from L_m by substituting a for x, and $Q_{(a)m,s}$ is obtained from $Q_{m,s}$ by the same substitution. As in the algebraic case we can form inner and outer products and each of these must be a formal covariant. We shall presently exhibit a system of covariants of the formal linear binary form modulo 3, but we shall first consider how Glenn†

* *Amer. Journ. of Maths.* vol. 36, pp. 449–460 (1914).

† *Bulletin Amer. Math. Soc.* vol. 21, p. 173 (1914–15).

found this system. He was the first to use modular operators. Since $x_1^{p^{nt}}, \ldots, x_m^{p^{nt}}$ are cogredient with $x_1, \ldots, x_m$,

$$x_1^{p^{nt}} \frac{\partial}{\partial x_1} + x_2^{p^{nt}} \frac{\partial}{\partial x_2} + \ldots + x_m^{p^{nt}} \frac{\partial}{\partial x_m} \equiv E_m^{(t)}$$

is an invariant operator. We shall call it the **MODULAR POLAR.** Similarly we have the **MODULAR ARONHOLD OPERATOR**

$$a_0^{p^{nt}} \frac{\partial}{\partial a_0} + a_1^{p^{nt}} \frac{\partial}{\partial a_1} + \ldots + a_\mu^{p^{nt}} \frac{\partial}{\partial a_\mu} \equiv F_\mu^{(t)}.$$

Again, if we have two m-ary quantics whose coefficients are $a_0, \ldots, a_\mu$ and $b_0, \ldots, b_\mu$ respectively, then we have two modular Aronhold operators

$$a_0^{p^{nt}} \frac{\partial}{\partial b_0} + a_1^{p^{nt}} \frac{\partial}{\partial b_1} + \ldots + a_\mu^{p^{nt}} \frac{\partial}{\partial b_\mu} \equiv H_{\mu, b}^{(a, t)};$$

$$b_0^{p^{nt}} \frac{\partial}{\partial a_0} + b_1^{p^{nt}} \frac{\partial}{\partial a_1} + \ldots + b_\mu^{p^{nt}} \frac{\partial}{\partial a_\mu} \equiv A_{\mu, a}^{(b, t)}.$$

Glenn also defines **MODULAR TRANSVECTANTS** for the binary case. For the general case if we have m functions $\phi_1, \phi_2, \ldots, \phi_m$ and if the degree of ϕ_i is t_i, then the rth transvectant with respect to $s_2, s_3, \ldots, s_m$ is obtained by operating r times with the Cayley operator Ω on

$$\phi_1(x),\ \phi_2(y),\ \ldots,\ \phi_m(z),$$

dividing by $\frac{\lfloor t_1}{\lfloor t_1 - r} \frac{\lfloor t_2}{\lfloor t_2 - r} \cdots \frac{\lfloor t_m}{\lfloor t_m - r}$, and then substituting $x_i^{p^{s_2}}$ for $y_i, \ldots, x_i^{p^{s_m}}$ for z_i. We shall write the expression thus obtained as

$$(\phi_1, \phi_2, \ldots, \phi_m)^r_{s_2 s_3 \ldots s_m}.$$

By means of such modular operators Glenn gives the following system for the formal covariants of the binary linear form for $p=3$. He does not show, however, that it is a fundamental or a full system. We append the symbolical representation

$$f = a_1 x_1 + a_2 x_2 = a_x,$$

$$E_2^{(1)} f = a_1 x_1^3 + a_2 x_2^3 = a_{x^3},$$

$$F_1^{(1)} f = a_1^3 x_1 + a_2^3 x_2 = a^3{}_x,$$

$$L_2 = x_1^3 x_2 - x_1 x_2^3 = (x^3 x),$$

$$Q_{2,1} = x_1^6 + x_1^4 x_2^2 + x_1^2 x_2^4 + x_2^6 = (x^9 x)/(x^3 x).$$

$$(L_2, f^4)_1^4 \ ||| \ a_1^3 a_2 - a_1 a_2^3 = (a^3 a),$$

$$(Q_{2,1}, f^6)_1^6 \ ||| \ a_1^6 + a_1^4 a_2^2 + a_1^2 a_2^4 + a_2^6 = (a^9 a)/(a^3 a),$$

$$(Q_{2,1}, f^3)_1^3 \ ||| \ a_2 (a_1^2 - a_2^2) x_1^3 - a_1^3 x_1^2 x_2 + a_2^3 x_1 x_2^2 - a_1 (a_2^2 - a_1^2) x_2^3$$
$$= [(a_{x^3})^2 (a^3 a) - (a^3{}_x)^2 (x^3 x)]/(a_x)^3.$$

Proceeding in this manner we can easily find a great many formal covariants, but so far we cannot say whether every covariant can be expressed in such a symbolical form. We do know, however, that L_m, $Q_{m,1}, \ldots, Q_{m,m-1}$ form a full system of universal covariants of m variables, so that interchanging α_s for x_s we have the finiteness of formal invariants modulo p of the m-ary linear form

$$a_1x_1 + \ldots + a_mx_m.$$

We can write this basis as

$$L_{(\alpha)m},\ Q_{(\alpha)m,1},\ \ldots,\ Q_{(\alpha),m,m-1}.$$

If the generalisation of Dickson's Theorem for several cogredient variables were known, then of course we should have a full system of formal invariants of a system of linear forms. Utilising Krathwohl's Theorem we have the following full system of formal invariants of a pair of binary linear forms

$$f = a_1x_1 + a_2x_2, \qquad g = b_1x_1 + b_2x_2.$$

$$L_a,\ L_b,\ Q_a,\ Q_b,\ M = \begin{vmatrix} a_1 & a_2 \\ b_1 & b_2 \end{vmatrix}, \qquad M_1 = \begin{vmatrix} a_1^p & a_2^p \\ b_1 & b_2 \end{vmatrix}, \qquad M_2 = \begin{vmatrix} a_1 & a_2 \\ b_1^p & b_2^p \end{vmatrix},$$

$$N_s = \frac{M_2^{s+1}L_a^{p-s-1} + (-)^s M_1^{p-s} L_b^s}{M^p} \qquad (1 \leqslant s \leqslant p-2).$$

§ 15. The use of Symbolical Operators.

We noticed in § 14 that the modular Aronhold operators were not symbolical operators. The symbolical operators

$$\alpha_1^p \frac{\partial}{\partial \alpha_1} + \alpha_2^p \frac{\partial}{\partial \alpha_2} + \ldots + \alpha_m^p \frac{\partial}{\partial \alpha_m} = \left(\alpha^p \,\middle|\, \frac{\partial}{\partial \alpha}\right)$$

are also invariant operators if used with certain restrictions. A function in the α's, β's, ... has no meaning unless it is symmetrical with respect to the α's, β's, ... and of the same degree in each. Thus the symbolic operators can be used provided the expression remains symmetrical in the α's, β's, ... and the total change of degree is the same for every symbol α, β, ... Thus $(\alpha\beta)^2$ is an invariant of $f = \alpha_x\beta_x$, $p = 3$: whence

$$\left(\alpha^3 \,\middle|\, \frac{\partial}{\partial \alpha}\right)\left(\beta^3 \,\middle|\, \frac{\partial}{\partial \beta}\right)(\alpha\beta)^2 \equiv -(\alpha^3\beta)(\beta^3\alpha) - (\alpha^3\alpha)(\beta^3\beta)$$

$$\equiv (\alpha\beta)^4 + (\alpha^3\alpha)(\beta^3\beta).$$

Thus we get the new invariant $(\alpha^3\alpha)(\beta^3\beta)$ since $(\alpha\beta)^4$ is an invariant.

Great care must be employed in using these operators to ensure that the factor p does not arise through their use. This factor may be trivial or it may not. We shall show an example where it is not trivial. For further use we shall give some symbolical formal invariants of the binary quadratic $p=2$:

$$f = a_0 x_1^2 + a_1 x_1 x_2 + a_2 x_2^2 = \alpha_x \beta_x,$$

$$(\alpha\beta) \mathrel{|||} a_1, \qquad (\alpha^2\alpha)(\beta^2\beta) \mathrel{|||} a_0 a_2 (a_0 + a_1 + a_2),$$

$$(\alpha^2\beta)^3(\beta^2\beta) + (\beta^2\alpha)^3(\alpha^2\alpha) \mathrel{|||} a_1^3(a_0+a_2)(a_1^2 + a_0a_2 + a_0a_1 + a_1a_2),$$

$$(\alpha^4\alpha)(\beta^4\beta)/(\alpha^2\alpha)(\beta^2\beta) \mathrel{|||} a_0^2 + a_1^2 + a_2^2 + a_0a_1 + a_1a_2 + a_2a_0.$$

That symbolical representation of covariants can be very complicated is seen by the following example of a covariant of f:

$$(a_0^2 + a_1^2 - a_0a_2)\,x_1^2 - a_1(a_0+a_2)\,x_1x_2 + (a_2^2 + a_1^2 - a_0a_2)\,x_2^2$$

$$\mathrel{|||} \frac{(\alpha\alpha^9)}{(\alpha\alpha^3)}(x_{\beta^3})^2 + \frac{(\beta\beta^9)}{(\beta\beta^3)}(x_{\alpha^3})^2 + [(\alpha^3\beta)^2(\alpha\beta)(\alpha\alpha^3) - (\alpha\beta^3)(\alpha\alpha^9)]\frac{\alpha_x\beta_x}{(\alpha\alpha^3)^2}.$$

It is comparatively easy to write down many symbolical formal covariants, but it is not always easy to find what their non-symbolical representations are. A good insight into the structure of such symbolical covariants is obtained by considering a few examples. To lessen the work entailed we can use the following method of abbreviation for the binary case:

We write $\quad \alpha_1^3\alpha_2^2\beta_1^4\beta_2$ as $3\,|\,4$; $\quad \boxed{5 \mid 5}$

$\alpha_1^2\alpha_2^2\beta_1^3\beta_2 - \alpha_1^2\alpha_2^2\beta_1\beta_2^3$ as $2\,|\,(3-1)$; $\quad \boxed{4 \mid 4}$

$\alpha_1^3\alpha_2^2\beta_1^5\gamma_2^5 + 2\alpha_1^3\alpha_2^2\beta_2^5\gamma_1^5$ as $3\,|\,5\,|\,0 + 2\;3\,|\,0\,|\,5$; $\quad \boxed{5 \mid 5 \mid 5}$

etc., that is, we simply write the indices of $\alpha_1, \beta_1, \gamma_1, \ldots$, with strokes between. The coefficients remain full size while the indices remain their own size. The figures in the oblongs at the side show how many α's, β's, etc. appear in each term. As an example of this notation we shall show that it is not in general permissible to divide by the modulus, even though the modulus appears only as a factor due to an operator. Let

$$f = a_0x_1^2 + a_1x_1x_2 + a_2x_2^2 = \alpha_x\beta_x \text{ and } p = 3.$$

Then $\qquad (\alpha\beta)^2 = \alpha_1^2\beta_2^2 - 2\alpha_1\alpha_2\beta_1\beta_2 + \alpha_2^2\beta_1^2.$

Now

$$\left[\left(\beta^3 \,|\, \frac{\partial}{\partial\beta}\right)\left(\beta^3 \,|\, \frac{\partial}{\partial\beta}\right)\left(\alpha^3 \,|\, \frac{\partial}{\partial\beta}\right) + \left(\alpha^3 \,|\, \frac{\partial}{\partial\alpha}\right)\left(\alpha^3 \,|\, \frac{\partial}{\partial\alpha}\right)\left(\beta^3 \,|\, \frac{\partial}{\partial\alpha}\right)\right](\alpha\beta)^2$$

$$= 6\,[\alpha_1^5\beta_1^3\beta_2^2 - \alpha_1^5\beta_1\beta_2^4 - \alpha_1^4\alpha_2\beta_2^5 + \alpha_1^3\alpha_2^2\beta_1^5 + \alpha_1^2\alpha_2^3\beta_2^5 - \alpha_1\alpha_2^4\beta_1^5$$
$$- \alpha_2^5\beta_1^4 + \alpha_2^5\beta_1^2\beta_2^3] \quad \ldots\ldots(15{\cdot}1)$$

$$= 6\,[5\,|\,(3-1) + 4\,|\,(-0) + 3\,|\,5 + 2\,|\,0 + 1\,|\,(-5) + 0\,|\,(-4+2)]; \quad \boxed{5 \mid 5}$$

this last must be a formal invariant modulo 3. If it were permissible to divide by 3 here, then would

$$5\,|\,(3-1)+4\,|\,(-0)+3\,|\,5+2\,|\,0+1\,|\,(-5)+0\,|\,(-4+2)\,; \qquad \boxed{5 \mid 5}$$

be an invariant. Call this K.

Now
$$(\alpha^3 a) \text{ is } 3\,|\,0-1\,|\,0\,; \qquad \boxed{4 \mid 0}$$
$$(\alpha\beta^3) \text{ is } 1\,|\,0-0\,|\,3\,; \qquad \boxed{1 \mid 3}$$
$$(\alpha\beta^3)^2 \text{ is } 2\,|\,0-2\,1\,|\,3+0\,|\,6\,; \qquad \boxed{2 \mid 6}$$
$$= 2\,|\,0+1\,|\,3+0\,|\,6\,; \qquad \boxed{2 \mid 6}\,.$$

Hence
$$(\alpha^3 a)(\alpha\beta^3)^2 = 5\,|\,0+4\,|\,3+3\,|\,(6-0)+2\,|\,(-3)+1\,(-6)\,; \qquad \boxed{6 \mid 6}$$

so that $(\alpha^3 a)(\alpha\beta^3)^2 - (\beta^3\beta)(\beta\alpha^3)^2 = L$, say, will be an invariant and

$$L = 6\,|\,(-3+1)+5\,|\,0+4\,|\,3+3\,|\,(6-4+2-0)$$
$$+2\,|\,(-3)+1\,|\,(-6)+0\,|\,(-5+3)\,; \qquad \boxed{6 \mid 6}\,.$$

We divide this by
$$(\alpha\beta) = 1\,|\,0-0\,|\,1\,; \qquad \boxed{1 \mid 1}$$

$$\begin{array}{l|l}
 & 5|(-3+1) \quad +4|(-4+2+0)+3|(-5-3+1) \quad +2|(4-2-0) \quad +1|(5+3-1) \quad +0|(4-2) \\ \hline
\underline{1|0-0|1} & 6|(-3+1)+5|0 \quad +4|3 \quad +3|(6-4+2-0)+2|(-3) \quad +1|(-6) \quad +0|(-5+3) \\
 & 5|(4-2) \quad 4|(5-3-1) \quad 3|(6+4-2) \quad 2|(-5+3+1) \quad 1|(-6-4+2)\,0|(-5+3) \\ \hline
 & 5|(-4+2+0)\,4|(-5-3+1) \quad 3|(4-2-0) \quad 2|(+5+3-1) \quad 1|(4-2) \quad \ldots
\end{array} ; \quad \boxed{5 \mid 5},$$

so that
$$K + \frac{L}{(\alpha\beta)} = 4\,|\,(-4+2)+3\,|\,(-3+1)+2\,|\,(4-2)+1\,|\,(3-1)\,; \qquad \boxed{5 \mid 5}\,.$$

Now
$$(\alpha^3 a)(\beta^3\beta) = (3\,|\,0-1\,|\,0)\,(0\,|\,3-0\,|\,1)\,; \qquad \boxed{4 \mid 4}$$
$$= 3\,|\,(3-1)+1\,|\,(-3+1)\,; \qquad \boxed{4 \mid 4}\,.$$

So that
$$\frac{K + L/(\alpha\beta)}{-(\alpha^3 a)(\beta^3\beta)} = 1\,|\,1+0\,|\,0\,; \qquad \boxed{1 \mid 1}$$
$$= a_1\beta_1 + a_2\beta_2$$
$$= a_0 + a_2.$$

Thus if K is an invariant so is $a_0 + a_2$, but $a_0 + a_2$ is not invariant under the transformation of matrix $\begin{bmatrix} 1 & 1 \\ \cdot & 1 \end{bmatrix}$. Therefore K is not an invariant; hence we may not divide by 3 in equation (15·1). This shows that the operator $\left(\alpha^3 \,\middle|\, \frac{\partial}{\partial a}\right)$ operating on α^3 is an invariant operator only in virtue of the factor 3 produced.

§16. Annihilators of Formal Invariants*.

In this paragraph we shall consider a method utilised by Glenn and Dickson for finding modular invariants in special cases. The modular

* *Amer. Journ. of Maths.* vol. 37, p. 75 (1915).

annihilator differs somewhat from the algebraic counterpart. The theory will be best illustrated by a simple example. We wish to find formal invariants of degree 4 of the binary quadratic form modulo 3. Let

$$f = a_0 x_1^2 + a_1 x_1 x_2 + a_2 x_2^2,$$

and let $$x_1 = \bar{x}_1 + t\bar{x}_2, \quad x_2 = \bar{x}_2.$$

Then $$\bar{f} = \bar{a}_0 \bar{x}_1^2 + \bar{a}_1 \bar{x}_1 \bar{x}_2 + \bar{a}_2 \bar{x}_2^2,$$

where
$$\begin{aligned} \bar{a}_0 &= a_0, \\ \bar{a}_1 &= a_1 + 2ta_0, \\ \bar{a}_2 &= a_2 + ta_1 + t^2 a_0. \end{aligned}$$

Let $\phi(a_0 a_1 a_2)$ be an invariant and let $\frac{\partial\phi}{\partial a_i}$ be written ϕ_{a_i}, then

$$\begin{aligned} &\phi(\bar{a}_0 \bar{a}_1 \bar{a}_2) \\ &= \phi(a_0 a_1 a_2) + 2ta_0 \phi_{a_1} + (ta_1 + t^2 a_0)\phi_{a_2} \\ &+ \frac{1}{\underline{|2}}[4t^2 a_0^2 \phi_{a_1^2} + 4ta_0(ta_1 + t^2 a_0)\phi_{a_1 a_2} + (ta_1 + t^2 a_0)^2 \phi_{a_2^2}] \\ &+ \frac{1}{\underline{|3}}[8t^3 a_0^3 \phi_{a_1^3} + 12t^2 a_0^2 (ta_1 + t^2 a_0)\phi_{a_1^2 a_2} + 6ta_0(ta_1 + t^2 a_0)^2 \phi_{a_1 a_2^2} \\ &+ (ta_1 + t^2 a_0)^3 \phi_{a_2^3}] + \frac{1}{\underline{|4}}[16t^4 a_0^4 \phi_{a_1^4} \\ &+ 32t^3 a_0^3 (ta_1 + t^2 a_0)\phi_{a_1^3 a_2} + 24t^2 a_0^2 (ta_1 + t^2 a_0)^2 \phi_{a_1^2 a_2^2} \\ &+ 8ta_0(ta_1 + t^2 a_0)^3 \phi_{a_1 a_2^3} + (ta_1 + t^2 a_0)^4 \phi_{a_2^4}] \quad \ldots\ldots\ldots\ldots\ldots\ldots (16{\cdot}1), \end{aligned}$$

in which we need not go farther than the fourth derivatives since the invariant is only of the fourth degree.

Now we can write equation (16·1) as

$$\phi(\bar{a}_0 \bar{a}_1 \bar{a}_2) - \phi(a_0 a_1 a_2) = t\delta_1\phi + t^2\delta_2\phi + \ldots + t^8\delta_8\phi.$$

Equations (11·1) show that ϕ must be of the form

$$\begin{aligned} \phi(a) = {} & A_1^{(0)} a_0^4 + A_1^{(2)} a_0^3 a_2 + A_2^{(2)} a_0^2 a_1^2 + A_1^{(4)} a_0^2 a_2^2 \\ & + A_2^{(4)} a_0 a_1^2 a_2 + A_3^{(4)} a_1^4 + A_1^{(6)} a_0 a_2^3 + A_2^{(6)} a_1^2 a_2^2 + A_1^{(8)} a_2^4, \end{aligned}$$

where the A's are undetermined constants. Hence

$$\begin{aligned} \delta_1\phi &= 2a_0\phi_{a_1} + a_1\phi_{a_2}, \\ \delta_3\phi &= 2a_0^2\phi_{a_1 a_2} + a_1 a_0 \phi_{a_2^2} + \tfrac{4}{3} a_0^3 \phi_{a_1^3} + 2a_0^2 a_1 \phi_{a_1^2 a_2} \\ &\qquad + a_0 a_1^2 \phi_{a_1 a_2^2} + \tfrac{1}{6} a_1^3 \phi_{a_2^3}, \\ \delta_5\phi &= a_0^3 \phi_{a_1 a_2^2} + \tfrac{1}{2} a_1 a_0^2 \phi_{a_2^3} + \tfrac{4}{3} a_0^4 \phi_{a_1^3 a_2} \\ &\qquad + 2a_0^3 a_1 \phi_{a_1^2 a_2^2} + a_1^2 a_0^2 \phi_{a_1 a_2^3} + \tfrac{1}{6} a_1^3 a_0 \phi_{a_2^4}, \\ \delta_7\phi &= \tfrac{1}{3} a_0^4 \phi_{a_1 a_2^3} + \tfrac{1}{6} a_0^3 a_1 \phi_{a_2^4}. \end{aligned}$$

We notice also that whenever the modulus 3 appears in the denominator, it will also occur in the numerator. We are still at liberty to cancel out these 3's as we have not yet introduced into our argument the fact that there is a modulus. We have so far proceeded exactly as in the algebraic case except for the fact that in the algebraic case the invariants are isobaric and hence we would have to set several of the A's equal to zero. In the formal case modulo 3, $\mathfrak{t} \parallel \mathfrak{t}^3 \parallel \mathfrak{t}^5 \parallel \mathfrak{t}^7$, and hence we obtain formal invariants if we put $\delta_1 + \delta_3 + \delta_5 + \delta_7 = 0$. This gives the following relations:

$$A_2^{(4)} - A_3^{(4)} + A_1^{(6)} \mathrel{|||} 0, \qquad A_1^{(4)} \mathrel{|||} A_2^{(4)},$$

$$A_2^{(6)} \mathrel{|||} A_1^{(8)} \mathrel{|||} 0, \qquad A_1^{(2)} + A_2^{(2)} + A_1^{(4)} - A_3^{(4)} \mathrel{|||} 0,$$

and since ϕ must be invariant under the substitution $x_1 = \bar{x}_2,\ x_2 = -\bar{x}_1$, it is invariant if we interchange a_0 with a_2 and $-a_1$ with a_1, so that

$$A_1^{(0)} \mathrel{|||} A_2^{(2)} \mathrel{|||} 0, \qquad A_1^{(2)} \mathrel{|||} A_1^{(6)}.$$

Hence $$\phi(a) \mathrel{|||} A_1^{(4)}(a_0^2 a_2^2 + a_0 a_1^2 a_2 - a_0^3 a_2 - a_0 a_2^3) + A_3^{(4)}(a_0^3 a_2 + a_1^4 + a_0 a_2^3).$$

Hence we have two linearly independent invariants of degree 4 of the binary quadratic, mod 3,

$$I \equiv a_0^2 a_2^2 + a_0 a_1^2 a_2 - a_0^3 a_2 - a_0 a_2^3, \qquad J \equiv a_0^3 a_2 + a_1^4 + a_0 a_2^3.$$

Glenn points out that $I + J \mathrel{|||} D^2$ where D is the discriminant of f, so that either I or J is reducible.

This method is somewhat cumbersome but it leads to all invariants.

By a different method Glenn found a full system of the binary cubic mod 2 and the binary quadratic mod 3. His method is somewhat long and will not be included here, but the procedure will be indicated. As in the algebraic theory, the coefficient of the highest power of x_1 is called the leader and the leader must be a seminvariant. We can always remove factors L so that a covariant can be taken generally as having the highest power of x_1 equal to the order of the covariant. If we have a fundamental system of seminvariants, then we try to find a fundamental system of invariants. If there be two covariants of the same order with the same leader, then their difference has the factor L, and we can reduce so that it is sufficient to consider only one covariant of a given order with a given leader. We would refer the reader to the original papers for further details*.

§ 17. Dickson's Method for Formal Covariants.

Let $f_1(x_1, \dots, x_m), \dots, f_t(x_1, \dots, x_m)$ be forms of total degrees $s_1, \dots, s_t$ in the independent variables $x_1, \dots, x_m$. Let g_i be the H.C.F. of q and s_i and let $q_i = q/g_i$. Then for any non-zero element ρ, of the $GF[p^n]$,

$$\rho^{s_i q_i} = (\rho^q)^{s_i/g_i} \parallel 1,$$

* Dickson, *Trans. Amer. Math. Soc.* vol. 14, p. 299 (1913) and Glenn, *ibid.* vol. 20, p. 154 (1919).

and hence $$[f_i(\rho x_1, \ldots, \rho x_m)]^{q_i} \Vert [f_i(x_1, \ldots, x_m)]^{q_i}.$$

Thus $f_i^{q_i}$ has a definite value at every point whose coordinates belong to $GF[p^n]$. We have seen in § 4 that all such points are conjugate, that is they are permuted by transformations of the group; hence the values $f_i^{q_i}$ also will be permuted by transformations of the group. Now let $\phi(f_1^{q_1}, \ldots, f_t^{q_t})$ be any function of $f_1^{q_1}, \ldots, f_t^{q_t}$; then ϕ will take different values $\phi_1, \phi_2, \ldots, \phi_P$ for the different points where there are P points with coordinates from the $GF(p^n)$. Any symmetric function of the P quantities $\phi_1, \ldots, \phi_P$ must be a formal covariant, since the effect of any transformation is merely to permute the different points.

As an example we consider the binary quadratic modulo 2

$$f = a_0x_1^2 + a_1x_1x_2 + a_2x_2^2.$$

The only points with coordinates from the $GF[2]$ are (1, 0), (0, 1), (1, 1), for which f takes the values a_0, a_2, $a_0 + a_1 + a_2$. Any symmetric function of these three must be a formal invariant. We easily infer that a_1, $a_0^2 + a_2^2 + a_0a_2 + a_0a_1 + a_1a_2$ and $a_0a_2(a_0 + a_1 + a_2)$ are formal invariants. Other examples are given by Dickson*.

We can also find certain covariants by an extension of this method. Let $y_1, y_2, \ldots, y_m$ be a set cogredient with $x_1, x_2, \ldots, x_m$; then

$$\phi = \begin{vmatrix} y_1 & y_2 & \cdots & y_m \\ y_1^p & y_2^p & \cdots & y_m^p \\ \cdots & \cdots & \cdots & \cdots \\ y_1^{p^{m-2}} & & \cdots & y_m^{p^{m-2}} \\ x_1 & & \cdots & x_m \end{vmatrix}$$

is an invariant which we can use as an auxilliary ground form. Now, if $f(a, x)$ be an m-ary ground form, then a simultaneous invariant of $f(a, x)$ and ϕ obtained by Dickson's method will be a formal covariant of $f(a, y)$.

A modification of these methods can be used in simple cases. If $p = 2$, we see from § 7 that every function which is invariant under

$$T_1 \quad \begin{array}{ll} x_1 = \bar{x}_1 + \bar{x}_2 & \\ x_i = \bar{x}_i & (i \neq 1) \end{array}$$

and under

$$T_3 \quad \begin{array}{ll} x_1 = \bar{x}_j & \\ x_j = \bar{x}_1 & (k \neq j,\ k \neq 1) \\ x_k = \bar{x}_k & \end{array}$$

* *Trans. Amer. Math. Soc.* vol. 15, pp. 499 ff. (1914).

is a formal covariant. Consider the case $p = m = 2$, $n = 1$ of the quadratic

$$f = a_0 x_1^2 + a_1 x_1 x_2 + a_2 x_2^2.$$

Under T_1 f becomes $\bar{f}$ where

$$\bar{a}_0 = a_0,$$
$$\bar{a}_1 = 2a_0 + a_1 \equiv a_1,$$
$$\bar{a}_2 = a_0 + a_1 + a_2.$$

A function of the a's which is invariant under T_1 is called a SEMINVARIANT. It is easy to see that

$$a_0,\ a_1,\ a_2 (a_0 + a_1 + a_2)$$

are seminvariants. Any function of these which is symmetrical with respect to a_0 and a_2 will be invariant under T_3 also, and will therefore be a formal invariant. Thus we have the following formal invariants

$$a_1,\ a_0 a_2 (a_0 + a_1 + a_2),\ a_0^2 + a_1 a_0 + a_0 a_2 + a_2 a_1 + a_2^2.$$

§ 18. Symbolical Representation of Pseudo-isobaric Formal Covariants.

If we consider the system of formal covariants of the binary linear form modulo 3 given in § 14, it seems at first sight as if every formal covariant could be expressed as a rational function of symbolical inner and outer products. This however is not the case. Let

$$f = a_0 x_1^3 + a_1 x_1^2 x_2 + a_2 x_1 x_2^2 + a_3 x_2^3.$$

Then $K = a_1 + a_2$ is known to be a formal invariant of f if $p = 2$, but it cannot be expressed as a rational function of symbolical inner and outer products. Miss Hazlett however noticed that if we write $f = \alpha_x \beta_x \gamma_x$,

then $$K^2 \equiv \Sigma\, (\alpha^2\gamma)(\beta^2\gamma) + (\alpha\beta)(\beta\gamma)(\gamma\alpha).$$

The remainder of this paragraph is an extension to the m-ary case of what Miss Hazlett* did for the binary case.

Let $\alpha^{p^{sn}} = \alpha^{(s)}$. Then $\alpha_1^{(s)}, \ldots, \alpha_m^{(s)}$ are simply a set of elements cogredient with the set $\alpha_1, \ldots, \alpha_m$. Let C, a pseudo-isobaric formal covariant, consist of terms C_i. The weights of the different terms must be congruent modulo q. Let the weight then of C_i with respect to x_1 be $w_1 + qW_{1i}$, where w_1 is the weight of the term of least weight with respect to x_1. Suppose first that $w_1 \neq 0$. Now C^{p^n} and therefore $\Sigma C_i^{p^n}$ are formal covariants. The weight of $C_i^{p^n}$ with respect to x_1 is $p^n (w_1 + qW_{1i})$. Now

* *Trans. Amer. Math. Soc.* vol. 14, pp. 300–304 (1913).

replace qW_{1i} symbols of the sort $\alpha_1^{p^n}, \beta_1^{p^n}, \ldots, \delta_1^{p^n}, x_2^{p^n}, \ldots, x_m^{p^n}$ by the corresponding symbols of the sort $\alpha_1^{(1)}, \beta_1^{(1)}, \ldots, \delta_1^{(1)}, x_2^{(1)}, \ldots, x_m^{(1)}$. Each replacement lowers the weight of the term by q, so that the weight of $C_i^{p^n}$ with respect to x_1 is now

$$p^n w_1 + q W_{1i}.$$

We do this for every term $C_i^{p^n}$. We make a further replacement of W_{1j} symbols in the terms where $w_1 > W_{1j}$. The weight of such terms $C_j^{p^n}$ is now $p^n w_1$. If there still exists terms $C_k^{p^n}$ for which $w_1 \leqslant W_{1k}$, we consider next the formal covariant $\Sigma C_i^{p^{2n}}$. The weight of the term $C_i^{p^{2n}}$ is now $p^{2n} w_1$ if $W_{1i} < w_1$, and is $(p^n w_1 + q W_{1i}) p^n$ if $W_{1i} \geqslant w_1$. We proceed as before, and if $W_{1i} < p^n w_1$ for all W_{1i}, then on making $p^n W_{1i}$ replacements of the types $\alpha_1^{p^n}$ by $\alpha_1^{(1)}$, $\alpha_1^{(1)p^n}$ by $\alpha_1^{(2)}, \ldots, x_m^{(1)p^n}$ by $x_m^{(2)}$, we find that the weight of every term is $p^{2n} w_1$. If $W_{1i} \geqslant p^n w_1$ for any W_{1i}, then we must go farther and consider $\Sigma C_i^{p^{3n}}$, and so on until we have $W_{1i} < p^{sn} w_1$ for all W_{1i}. Then we have that C is isobaric with respect to x_1 in the symbols $\alpha, \alpha^{(1)}, \alpha^{(2)}, \ldots, \alpha^{(s)}, \beta, \beta^{(1)}, \ldots, x^{(s)}$. We proceed similarly with respect to $x_2, \ldots, x_m$ and finally we have that $\Sigma C_i^{p^{sn}}$ is completely isobaric in the corresponding symbols. We therefore have the important result that *if C be a formal covariant then $C^{p^{ns}}$ for some s is congruent to an isobaric polynomial of the symbols* $\alpha, \alpha^{(1)}, \ldots, \alpha^{(s)}, \beta, \beta^{(1)}, \ldots, x^{(s)}$, i.e. *to a simultaneous congruent covariant of*

$$f(a, x), f(a^{p^n}, x), \ldots, f(a^{p^{sn}}, x), f(a, x^{p^n}), \ldots, f(a, x^{p^{sn}}) \quad \ldots\ldots(18\cdot1).$$

If $w_1 = 0$, we first multiply C by some algebraic invariant I whose weight is greater than zero and then the proof holds for IC, so that C is congruent to the quotient of two such simultaneous congruent covariants of the ground forms (18·1).

Thus if it be true that every congruent covariant can be represented symbolically it will also be true that every formal covariant can be represented symbolically in terms of the secondary symbols $\alpha^{(1)}, \alpha^{(2)}, \ldots, x^{(s)}$. As we have seen in § 10 however, this has been proved for the binary case only.

§ 19. Classes.

We must now consider the theory of classes and characteristic invariants. Dickson found this the best method for dealing with residual invariants. Now if the coefficients of the ground form f belong to $GF[p^n]$

there will be a finite number of possible f's. If two of these can be transformed into each other by a transformation of the group, then they are said to belong to the same CLASS. If they cannot be transformed into each other by a transformation of the group, then they belong to different classes. All the f's can thus be separated into classes such that any two f's in one class are transformable while any two f's of different classes are not transformable. We shall use as an illustration the binary quadratic

$$f \equiv ax^2 + bxy + cy^2,$$

where a, b, c belong to the $GF[2]$. The different possible types of f are evidently

$$\left.\begin{array}{ll} f_0 \equiv 0, & f_4 \equiv x^2, \\ f_1 \equiv xy, & f_5 \equiv x^2 + xy, \\ f_2 \equiv y^2, & f_6 \equiv x^2 + y^2, \\ f_3 \equiv xy + y^2, & f_7 \equiv x^2 + xy + y^2; \end{array}\right\} \quad \text{............(19·1)}$$

while the possible substitutions of the group are

$$\left.\begin{array}{lll} T_1 : \begin{bmatrix} 1 & . \\ . & 1 \end{bmatrix}, & T_2 : \begin{bmatrix} . & 1 \\ 1 & . \end{bmatrix}, & T_3 : \begin{bmatrix} 1 & 1 \\ 1 & . \end{bmatrix}, \\ T_4 : \begin{bmatrix} 1 & 1 \\ . & 1 \end{bmatrix}, & T_5 : \begin{bmatrix} 1 & . \\ 1 & 1 \end{bmatrix}, & T_6 : \begin{bmatrix} . & 1 \\ 1 & 1 \end{bmatrix}. \end{array}\right\} \quad \text{......(19·2)}$$

We tabulate the results obtained by transforming every f with every T in the following manner, where e.g. $T_4 f_3 = f_1$:

	f_0	f_1	f_2	f_3	f_4	f_5	f_6	f_7
T_1	f_0	f_1	f_2	f_3	f_4	f_5	f_6	f_7
T_2	f_0	f_1	f_4	f_5	f_2	f_3	f_6	f_7
T_3	f_0	f_5	f_4	f_1	f_6	f_3	f_2	f_7
T_4	f_0	f_3	f_2	f_1	f_6	f_5	f_4	f_7
T_5	f_0	f_5	f_6	f_3	f_4	f_1	f_2	f_7
T_6	f_0	f_3	f_6	f_5	f_2	f_1	f_4	f_7

......(19·3).

We therefore separate $f_0, \dots, f_7$ into the following four classes:

$$\left.\begin{array}{l} C_0 \text{ containing } f_0, \\ C_1 \text{ containing } f_1 f_3 f_5, \\ C_2 \text{ containing } f_2 f_4 f_6, \\ C_3 \text{ containing } f_7. \end{array}\right\} \quad \dots\dots\dots\dots\dots\dots (19{\cdot}4)$$

We can proceed in this manner in the general case even where we have more than one ground form. For example, Dickson gives as typical members of two m-ary linear forms the four classes

$$\begin{array}{lll} C_1; & x_1, & dx_2 \\ C_2; & x_1, & cx_1 \\ C_3; & 0, & x_1 \\ C_0; & 0, & 0. \end{array}$$

Classes under a group may be further separated into sub-classes by a sub-group of the group. Considering, then, these classes we see that we may expect a residual invariant to take different values for different classes. Indeed this gives us another definition of a residual invariant, namely a residual invariant is a function of the coefficients of a ground form such that it has the same value for two f's belonging to the same class, but may in certain cases take different values if the two f's do not belong to the same class.

§ 20. Characteristic Invariants*.

We now come to the important functions called CHARACTERISTIC INVARIANTS. The characteristic invariant I_j of a class C_j is such that it takes the value 1 for every f belonging to C_j but the value 0 for any other f. Suppose that any particular residual covariant V takes the values $v_0, \dots, v_{k-1}$ for the classes $C_0, \dots, C_{k-1}$, then obviously

$$V = v_0 I_0 + v_1 I_1 + \dots + v_{k-1} I_{k-1} \quad \dots\dots\dots\dots\dots (20{\cdot}1).$$

The total number of possible invariants of this sort is manifestly p^{nk}.

A set of invariants $R_1, \dots, R_l$ are said to CHARACTERISE the classes $C_1, \dots, C_k$ if they form a criterion for determining to which class a particular form belongs. Thus if an invariant R_1 takes a different value for each class, then it characterises the classes. Suppose however that it takes a different value for all classes with the exception of C_i and C_j for which it takes the same value; then it will not distinguish between C_i and C_j and it does not therefore characterise the classes. If however

* *Amer. Jour. of Maths.* vol. 31, pp. 349–366 (1909).

we have a second invariant R_2 which takes different values for the classes C_i and C_j, then R_1 and R_2 together characterise the classes.

From definition it is obvious that the characteristic invariants characterise the classes. We shall now consider a few of the properties of the characteristic invariants.

THEOREM. *Any residual invariant can be expressed in one, and only one, way as a homogeneous linear function of the characteristic invariants with coefficients of the field.*

PROOF. If it were possible to express the invariant in two ways, then we should have

$$V = v_0 I_0 + v_1 I_1 + \dots + v_{k-1} I_{k-1}$$
$$= u_0 I_0 + u_1 I_1 + \dots + u_{k-1} I_{k-1} \quad \dots\dots\dots\dots (20{\cdot}2),$$

whence $$0 = (v_0 - u_0) I_0 + (v_1 - u_1) I_1 + \dots + (v_{k-1} - u_{k-1}) I_{k-1} \dots (20{\cdot}3).$$

Now for a class C_i we have $I_i = 1$ and $I_j = 0$ $(j \neq i)$, so that $0 = v_i - u_i$ or $v_i = u_i$. We get this result for every value of i and thus the theorem is proved. We also see from this theorem that the number of homogeneously linear independent residual invariants is equal to the number of non-equivalent classes. We notice however the non-homogeneous relation

$$1 = I_0 + I_1 + \dots + I_{k-1} \quad \dots\dots\dots\dots\dots\dots (20{\cdot}4).$$

From the definition of characteristic invariants we obtain the following relations:

$$I_i^2 = I_i, \quad I_i I_j = 0 \qquad (i \neq j) \quad \dots\dots\dots (20{\cdot}5).$$

By means of these, any relation between the characteristic invariants can be reduced to a linear one which can be made homogeneous by means of (20·4). Thus any relation between the I's can be represented as

$$\Sigma c_i I_i = 0 \quad \dots\dots\dots\dots\dots\dots\dots\dots\dots (20{\cdot}6),$$

whence every c_i is zero. Thus (20·4) and (20·5) are the only relations between the characteristic invariants. Dickson* has given two rather artificial formulae for these characteristic invariants in the general case. If a system of forms has coefficients $c_1, \dots, c_s$, we have

$$I_k = \Sigma \prod_{i=1}^{s} \{1 - (c_i - c_i^{(k)})^q\} \quad \dots\dots\dots\dots\dots\dots (20{\cdot}7),$$

where the sum extends over all sets of coefficients $c_1^{(k)}, \dots, c_s^{(k)}$, of the

* *Madison Colloquium Lectures*, p. 13 and *Amer. Journ. of Maths.* vol. 31, p. 339 (1909).

various systems of the class C_k. The second formula is obtained by making use of Lagrange's interpolation formula

$$I_k = \sum_{c_1^{(k)}\dots c_s^{(k)}} \prod_{j=1}^{s} \frac{c_j^{p^n}-c_j}{c_j^{(k)}-c_j} \quad \dots\dots\dots\dots\dots\dots(20{\cdot}8).$$

Both of these expressions give

$$I_0 = \prod_{i=1}^{s}(1-c_i^{q}) \quad \dots\dots\dots\dots\dots\dots\dots\dots(20{\cdot}9).$$

These results are not of much practical value, but they prove the existence of the characteristic invariants.

§ 21. Syzygies*.

Let $L_1, \dots, L_\sigma$ be a full system of residual invariants, then every characteristic invariant must be a function of $L_1, \dots, L_\sigma$. Let $I_i = \phi_i(L_1, \dots, L_\sigma)$, then since $I_i I_j \parallel 0$ $(i \neq j)$ then $\phi_i \phi_j \parallel 0$ gives a relation between $L_1, \dots, L_\sigma$ and is therefore a syzygy. Similarly every L_i is expressible as a linear homogeneous function of the characteristic invariants so that every syzygy can be represented as a function G of the characteristic invariants, which is equal to zero; thus

$$G(I_0, \dots, I_{k-1}) \parallel 0.$$

But we have already shown that the only relations between the I's are (20·4) and (20·5). Therefore $G \parallel 0$ must be a result of combining (20·4) and (20·5).

Now suppose $K_0, \dots, K_{k-1}$ is another fundamental system. Since $K_h \parallel \sum_{\lambda=0}^{k-1} a_{\lambda h} I_\lambda$ and $K_j \parallel \sum_{\mu=0}^{k-1} a_{\mu j} I_\mu$, then

$$K_h K_j \parallel \sum_{\lambda=0}^{k-1} a_{\lambda h} a_{\lambda j} I_\lambda \parallel \sum_{\nu=0}^{k-1} c_{\nu h j} K_\nu \quad \dots\dots\dots\dots(21{\cdot}1),$$

where

$$c_{\nu j h} = c_{\nu h j} = \sum_{\lambda=0}^{k-1} a_{\lambda h} a_{\lambda j} A_{\lambda \nu} \quad \dots\dots\dots\dots\dots\dots(21{\cdot}2),$$

if $A_{\lambda\nu}$ be the cofactor of $a_{\lambda\nu}$ in $|a|$ divided by $|a|$.

Thus every product of the K's can be represented as a linear homogeneous expression in the K's themselves. If now

$$G(K_0, K_1, \dots, K_{k-1}) \parallel 0$$

is a syzygy between the K's, then by means of (21·1) we can reduce the syzygy to a linear relation between the K's, or else both sides vanish. Suppose then that this relation is expressed by

$$\beta_0 K_0 + \beta_1 K_1 + \dots + \beta_{k-1} K_{k-1} \parallel \beta \qquad (\beta \nparallel 0) \quad \dots(21{\cdot}3).$$

* From Prof. Weitzenböck's notes.

But $K_\nu \parallel \sum_{\lambda=0}^{k-1} a_{\lambda\nu} I_\lambda$, and so we have

$$\sum_{\nu=0}^{k-1} \sum_{\lambda=0}^{k-1} \beta_\nu a_{\lambda\nu} I_\lambda \parallel \beta.$$

Therefore

$$\beta_0 a_{\lambda 0} + \beta_1 a_{\lambda 1} + \dots + \beta_{k-1} a_{\lambda, k-1} \parallel \beta \qquad (\lambda = 0, 1, \dots, k-1)$$

so that $$\beta_\nu \parallel \beta \sum_\lambda A_{\lambda\nu} \parallel \beta (A_{0\nu} + A_{1\nu} + \dots + A_{k-1,\nu}),$$

and substituting in (21·3),

$$\sum_{\nu=0}^{\nu=k-1} (A_{0\nu} + A_{1\nu} + \dots + A_{k-1,\nu}) K_\nu \parallel 1 \quad \text{.........(21·4)}.$$

(21·4) and (21·1) we call ground syzygies. These correspond to (20·4) and (20·5) in the case of characteristic invariants.

THEOREM. *If a residual invariant J takes the value J_i for the class C_i $(i = 0, \dots, k-1)$, and if g of these values J_i are different when $2 \leqslant g \leqslant p$, then J satisfies a congruence of the gth degree, but satisfies no congruence of lower degree.*

PROOF. From $J \parallel J_0 I_0 + J_1 I_1 + \dots + J_{k-1} I_{k-1}$ and (20·5) we have $JI_\rho \parallel J_\rho I_\rho$ or $(J - J_\rho) I_\rho \parallel 0$. Give ρ the values $0, \dots, k-1$, then since one I_ρ must be congruent to 1, one factor $(J - J_\rho)$ must be congruent to 0, and therefore

$$(J - J_0)(J - J_1) \dots (J - J_{k-1}) \parallel 0.$$

Let $J_{\rho_1}, \dots, J_{\rho_g}$ be the g different values of J, then obviously

$$(J - J_{\rho_1})(J - J_{\rho_2}) \dots (J - J_{\rho_g}) \parallel 0$$

is a congruence satisfied by J and is of degree g. If J satisfied a congruence of degree $s < g$, then the congruence could not have g different roots $J_{\rho_1}, \dots, J_{\rho_g}$ and so there is no congruence of degree less than g.

§ 22. Residual Covariants.

Every formal covariant is a residual covariant. We can of course reduce the non-symbolical coefficients of a formal covariant by Fermat's Theorem, so that the function remains a residual covariant but is no longer a formal covariant. The residual covariant thus obtained turns out in many cases to be zero. The most obvious method then of finding residual covariants is first to find formal covariants by any method suitable and then to reduce the non-symbolical coefficients by Fermat's Theorem.

In the case of residual invariants we can use the following theorem to find whether we have a full system or not. *If the residual invariants $R_1, \ldots, R_l$ completely characterise the non-equivalent classes, then they form a full system of residual invariants.* Let $c_1, \ldots, c_s$ be the coefficients of the forms in the system S. For the resulting p^{sn} sets of values of the c's let $R_1, \ldots, R_l$ take the distinct sets of values

$$R_{1(i)}, R_{2(i)}, \ldots, R_{l(i)} \qquad (i = 0, \ldots, N-1).$$

There are thus N non-equivalent classes in the system S; and by hypothesis the ith class is uniquely defined by the values $R_{1(i)}, \ldots, R_{l(i)}$. If an invariant Q takes the value Q_i for the class C_i, then

$$Q \parallel \sum_{i=0}^{N-1} Q_i \{1-(R_1-R_{1(i)})^q\} \ldots \{1-(R_l-R_{l(i)})^q\}.$$

Hence any invariant can be expressed as a polynomial in the $R_1, \ldots, R_l$ with coefficients from the $GF[p^n]$. Thus the invariants $R_1, \ldots, R_l$ form a full system.

The converse of this theorem is also true, for if $R_1, \ldots, R_l$ are a full system every characteristic invariant must be a function of $R_1, \ldots, R_l$. Since the characteristic invariants characterise the classes, so also must $R_1, \ldots, R_l$.

In § 15 and § 17 we found by different methods that a_1 and $a_0^2 + a_2^2 + a_0a_2 + a_0a_1 + a_1a_2$ were formal invariants mod 2 of

$$a_0x_1^2 + a_1x_1x_2 + a_2x_2^2.$$

Therefore using Fermat's Theorem we find that b, $a+c+ab+bc+ca$ are residual invariants of the modular form $ax^2+bxy+cy^2$; moreover they characterise the classes, for they take the following values for the different classes:

	b	$a+c+ab+bc+ca$
C_0	0	0
C_1	1	0
C_2	0	1
C_3	1	1

Thus b and $a+c+ab+bc+ca$ form a full system of invariants. As will be shown in a later paragraph, they also form a smallest full system.

From the theorem of the next paragraph we have a means of obtaining a full system of residual invariants if we know a full system of formal invariants. Miss Sanderson's Theorem has not, so far, been extended to covariants, and so at present there is no definite method of proving whether a given system of residual covariants is a full one or not. It seems likely however that a full system of residual covariants can be obtained from a full system of formal covariants.

Glenn* has given the following full system of formal covariants of $f = a_0x_1^2 + a_1x_1x_2 + a_2x_2^2$, where $p = 3$:

$$\Delta = a_1^2 - a_0a_2$$
$$J = a_0(a_0 - a_1 + a_2)(a_0 + a_1 + a_2)a_2$$
$$B = a_1(a_1^2 - a_0^2)(a_0 - a_2)(a_2^2 - a_1^2)$$
$$\Gamma = (a_0 + a_2)(-a_0 + a_1 + a_2)(-a_0 - a_1 + a_2)$$
$$L = x_1^3x_2 - x_1x_2^3$$
$$Q = x_1^6 + x_1^4x_2^2 + x_1^2x_2^4 + x_2^6$$
$$-C_1 = (a_0^2a_1 - a_1^3)x_1^2 + (a_2 - a_0)(a_1^2 + a_0a_2)x_1x_2 + (a_1^3 - a_1a_2^2)x_2^2$$
$$C_2 = (a_0^2 + a_1^2 - a_0a_2)x_1^2 - a_1(a_0 + a_2)x_1x_2 + (a_2^2 + a_1^2 - a_0a_2)x_2^2$$
$$C_4 = (a_2^2 + a_1^2 - a_0a_2)x_1^4 + a_1(a_0 + a_2)(x_1^3x_2 + x_1x_2^3) + (a_0^2 + a_1^2 - a_0a_2)x_2^4$$
$$f_4 = a_0x_1^4 - a_1(x_1^3x_2 + x_1x_2^3) + a_2x_2^4$$
$$f_6 = a_0x_1^6 + a_1x_1^3x_2^3 + a_2x_2^6$$
$$-\zeta_4 = (a_0^2a_1 - a_1^3)x_1^4 + (a_0 - a_2)(a_1^2 + a_0a_2)(x_1^3x_2 + x_1x_2^3) + (a_1^3 - a_1a_2^2)x_2^4$$
$$\zeta_6 = (a_0^2a_1 - a_1^3)x_1^6 - (a_0 - a_2)(a_1^2 + a_0a_2)x_1^3x_2^3 + (a_1^3 - a_1a_2^2)x_2^6$$
$$\phi_2 = a_0^3x_1^2 + a_1^3x_1x_2 + a_2^3x_2^2$$
$$\phi_4 = a_0^3x_1^4 - a_1^3(x_1^3x_2 + x_1x_2^3) + a_2^3x_2^4$$
$$\theta_2 = (a_0^3a_1 - a_0a_1^3)x_1^2 + (a_0a_2^3 - a_0^3a_2)x_1x_2 + (a_1^3a_2 - a_1a_2^3)x_2^2$$
$$\xi_2 = (a_1a_0^4 - a_0^2a_1^3)x_1^2 + (a_0^3a_1^2 - a_0^4a_2 - a_0a_1^4 + a_1^4a_2 - a_0^3a_2^2 + a_0^2a_2^3 - a_1^2a_2^3 + a_0a_2^4)x_1x_2 + (a_1^3a_2^2 - a_1a_2^4)x_2^2.$$

We now suppose that a_0, a_1, a_2 are elements of $GF[3]$; this gives at once

$$\phi_2 \parallel f, \qquad \phi_4 \parallel f_4,$$
$$B \parallel \zeta_2 \parallel \xi_2 \parallel 0.$$

We shall however expect to find relations amongst the remaining $f, \Delta, J, \Gamma, L, Q, C_1, C_2, C_4, f_4, f_6, \zeta_4, \zeta_6$, thus $f_6 \parallel f^3$, $-\zeta_6 \parallel C_1^3$.

Dickson† gives as a full system of residual covariants

$$f,\ \Delta,\ C_1,\ C_2, f_4,\ L,\ Q, \text{ and } q = (a_0 + a_2)(a_1^2 + a_0a_2 - 1).$$

Also
$$\begin{aligned}\Gamma &\parallel\!\mid (a_0 + a_2)(a_0^2 - a_1^2 + a_2^2 + a_0a_2)\\ &\parallel (a_0 + a_2)(a_1^2 + a_0a_2 - 1) + (a_0 + a_2)(a_0^2 + a_1^2 + a_2^2 + 1)\\ &\parallel q + 2a_0 + a_0a_1^2 + a_0a_2^2 + a_2a_0^2 + a_2a_1^2 + 2a_2\\ &\parallel q + (a_0 + a_2)(a_1^2 + a_0a_2 - 1)\\ &\parallel 2q.\end{aligned}$$

So we see that every residual covariant in this case can be obtained from

* *Trans. Amer. Math. Soc.* vol. 20, p. 154 (1919). † *Ibid* vol. 14, p. 310 (1913).

a formal covariant; the methods of proof that the above systems are full systems are not sufficiently general to be included here.

§ 23. Miss Sanderson's Theorem.

Miss Sanderson's Theorem* can be stated as follows: *To any residual invariant i of a system of forms under a group Γ with coefficients from the $GF[p^n]$, there corresponds a formal invariant I under Γ such that, for all sets of values in the field of the coefficients of the system of forms,*

$$I \parallel i.$$

We require the Lemma—*Let $a_1, \dots, a_r$ $(r > 1)$ be arbitrary variables and $g_1, \dots, g_r$ given elements of the $GF[p^n]$, $g_r \neq 0$, then the determinant*

$$N = \begin{vmatrix} a_1 & \dots & a_r \\ a_1^{p^n} & \dots & a_r^{p^n} \\ \dots & \dots & \dots \\ a_1^{p^{(r-1)n}} & \dots & a_r^{p^{(r-1)n}} \end{vmatrix} \qquad \dots\dots\dots\dots(23\cdot1)$$

is divisible in the field by the determinant

$$D = \begin{vmatrix} a_1 & \dots & a_r \\ a_1^{p^n} & \dots & a_r^{p^n} \\ \dots & \dots & \dots \\ a_1^{p^{(r-2)n}} & \dots & a_r^{p^{(r-2)n}} \\ g_1 & \dots & g_r \end{vmatrix} \qquad \dots\dots\dots\dots(23\cdot2)$$

and the quotient $Q = \dfrac{N}{D}$ has the properties

$$\left.\begin{array}{l} Q \neq 0 \text{ if } a_1 = g_1, \dots, a_r = g_r \\ Q = 0 \text{ if } a_1 = e_1, \dots, a_r = e_r \end{array}\right\} \qquad \dots\dots\dots(23\cdot3),$$

where $e_1, \dots, e_r$ are elements of the field not proportional to $g_1, \dots, g_r$.

We refer the reader to Miss Sanderson's paper for a proof of this lemma.

We consider a system S of forms $f_1, \dots, f_t$ in m variables with coefficients $a_1, \dots, a_r$. We separate all such systems into classes C_i. If S' is a particular system of forms $f_1', \dots, f_t'$ and if k is a constant, then we shall say that the system $kf_1', \dots, kf_t'$ is a MULTIPLE of S' and shall denote it by kS'. Now let c_i be a sub-set of C_i such that if s is in c_i then no multiple of s is in c_i, and such that every system S in C_i is a multiple of some system in c_i. Let ρ be a primitive root of the $GF[p^n]$, i.e. if $\rho^y = 1$, then y must be an integral multiple of $p^n - 1$. Let e_i be the

* *Trans. Amer. Math. Soc.* vol. 14, p. 490 (1913).

smallest exponent for which s' and $\rho^{e_i}s'$ are equivalent, where s' is any system in c_i. Then e_i is a factor of p^n-1 and

$$s',\ \rho^{e_i}s',\ \dots,\ \rho^{(d_i-1)e_i}s' \qquad (d_i e_i = p^n - 1) \quad \dots\dots\dots(23{\cdot}4)$$

are all contained in C_i. If S'' is a system in C_i not in the set (23·4), then $\rho^{e_i}S''$ will be in C_i. For S'' may be transformed into s', s' into $\rho^{e_i}s'$, $\rho^{e_i}s'$ into $\rho^{e_i}S''$. Moreover any system S''' in C_i is equal to ρ^{ke_i} times a system s''' in c_i, for we may write $S'''=\rho^{ke_i+l}s'''$ $(0 \leqslant l < e_i)$. Since $\rho^{p^n-1-ke_i}S'''$ is in C_i, $\rho^l s'''$ is in C_i. But e_i is the smallest exponent for which this is possible, hence $l=0$, and we can write

$$C_i = \sum_{k=1}^{d_i} c_i \rho^{ke_i},$$

where Σ denotes an aggregate, and $c_i\rho^{ke_i}$ is the set of systems obtained by multiplying each system in c_i by ρ^{ke_i}. If $e_i=1$, C_i will contain all the multiples of the systems in c_i. In general there are e_i different classes,

$$C_{il} = \sum_{k=1}^{d_i} c_i \rho^{ke_i+l} \qquad (l=0,\ 1,\ 2,\ \dots,\ e_i-1) \quad \dots\dots(23{\cdot}5),$$

formed on the sub-sets c_i, $c_i\rho$, ..., $c_i\rho^{e_i-1}$ respectively. Thus a complete list of the classes is given by C_{il} $(i=1,\ \dots,\ h;\ l=0,\ 1,\ \dots,\ e_i-1)$ and C_{00} which contains only the identically zero system.

We have seen in § 20 that all invariants of the system S of forms can be expressed in terms of characteristic invariants i_{kl}, which have the value 1 for the class C_{kl} and the value 0 for every other class. To prove our theorem it will be sufficient to construct formal invariants I_{kl} which reduce to i_{kl} if the coefficients are in the field. For this purpose we find it convenient to employ the invariants

$$J_i = \sum_{g_1 \dots g_r}^{c_i} [Q(g_1,\ \dots,\ g_r)]^{d_i} \qquad (i=1,\ \dots,\ h) \quad \dots(23{\cdot}6),$$

where the summation extends over the different sets of coefficients in c_i. When the variables are transformed, the coefficients a_1, ..., a_r undergo an induced transformation which is also linear with coefficients in $GF[p^n]$. J_i is a formal invariant, since the numerators are invariant apart from a factor which is the d_ith power of the determinant of the induced transformation and the denominators are permuted among themselves apart from the same factor.

Consider any particular denominator

$$\begin{vmatrix} a_1 & \dots & a_r \\ \dots & \dots & \dots \\ a_1^{p^{(r-2)n}} & \dots & a_r^{p^{(r-2)n}} \\ g_1' & \dots & g_r' \end{vmatrix}^{d_i} \quad \dots\dots\dots\dots\dots\dots(23{\cdot}7).$$

Since the $g_1', \ldots, g_r'$ are the coefficients of a system of c_i, this system is also in C_{i0}. After the transformation $a_i = \Sigma c_{ij} A_j$ this becomes

$$\begin{vmatrix} \Sigma c_{1j} A_j & \ldots \Sigma c_{rj} A_j \\ \ldots\ldots\ldots & \ldots\ldots\ldots \\ \Sigma c_{1j} A_j^{p^{(r-2)n}} & \ldots \Sigma c_{rj} A_j^{p^{(r-2)n}} \\ \Sigma c_{1j} G_j & \ldots \Sigma c_{rj} G_j \end{vmatrix}^{d_i} \quad \left(\Sigma \text{ means } \sum_{j=1}^{r}\right) \qquad \ldots\ldots(23\cdot8),$$

where the $G_1, \ldots, G_r$ are also the coefficients of a system in C_{i0} and hence may be written $\rho^{le_i} g_1'', \ldots, \rho^{le_i} g_r''$, where $g_1'', \ldots, g_r''$ are the coefficients of a system of c_i. Hence (23·8) becomes

$$\rho^{le_i d_i} | c_{ij} |^{d_i} \begin{vmatrix} A_1 & \ldots A_r \\ \ldots\ldots & \ldots\ldots \\ A_1^{p^{(r-2)n}} & \ldots A_r^{p^{(r-2)n}} \\ g_1'' & \ldots g_r'' \end{vmatrix}^{d_i}.$$

Since $\rho^{le_i d_i} = 1$, this denominator apart from the factor $| c_{ij} |^{d_i}$ occurs among the denominators in the sum (23·6), and the factor $| c_{ij} |^{d_i}$ is cancelled by the same factor which is brought into the numerator by the transformation: hence J_i is a formal invariant.

We shall denote "the value which J_i has for systems of the class C_{jk}" by $J_i(C_{jk})$. Then obviously by (23·3) and (23·6)

$$J_i(C_{00}) = 0, \qquad J_i(C_{kl}) = 0 \quad (k \neq i), \qquad J_i(C_{il}) \neq 0.$$

Since $J_i(C_{i0}) \neq 0$, we may put $J_i(C_{i0}) = \rho^l$ and it is easy to show that

$$J_i(C_{ik}) = \rho^{l+kd_i} \qquad (k = 0, \ldots, e_i - 1).$$

The theorem of § 20 implies that any invariant V can be written

$$V = \sum_{jk} v_{jk} I_{jk}, \text{ where } V(C_{jk}) = v_{jk}.$$

We can therefore determine the I_{ik} from

$$J_i^g = \sum_{k=0}^{e_i-1} \rho^{g(l+kd_i)} I_{ik} \qquad (g = 1, \ldots, e_i);$$

for the determinant of the coefficients can be shown to be non-singular. For C_{00} we have

$$I_{00} = 1 - \sum_{g_1 \ldots g_r}^{c} [Q(g_1, \ldots, g_r)]^q.$$

We now have a theoretical if not a very practical method of obtaining the formal invariants I_{jk} which reduce to the residual invariants i_{jk} of the class c_{jk}. The theorem is therefore proved.

Miss Sanderson's Theorem shows us that if we have a full system of formal invariants we can easily obtain the corresponding full system of residual invariants. For the binary linear form $f = a_1 x_1 + a_2 x_2$ mod 3 we have the full system of formal invariants

$$a_1{}^3 a_2 - a_1 a_2{}^3, \qquad a_1{}^6 + a_1{}^4 a_2{}^2 + a_1{}^2 a_2{}^4 + a_2{}^6,$$

which reduce to 0 and $a_1{}^2 + a_1{}^2 a_2{}^2 + a_2{}^2$ respectively when the a's belong to $GF[3]$, so that a full system of residual invariants of the binary linear form with coefficients from $GF[3]$ is given by

$$a_1{}^2 + a_1{}^2 a_2{}^2 + a_2{}^2.$$

A residual covariant does not take a definite value for each class and so we cannot write a residual covariant C as

$$C = I_0 c_0 + I_1 c_1, \dots, I_{k-1} c_{k-1}.$$

Miss Sanderson's Theorem therefore cannot be directly adapted for covariants. It seems likely, however, that every residual covariant can be obtained from a formal covariant; and in §22 we have seen that this is true in one case.

§ 24. A method of finding Characteristic Invariants.

THEOREM. *Let $K_0, \dots, K_{k-1}$ be k absolute rational integral residual invariants such that $a_{\lambda\rho}$ is the value of K_ρ for the class C_λ where there are k classes $C_0, \dots, C_{k-1}$; these k invariants will form a fundamental system if and only if $|a_{\lambda\rho}| \neq 0$.*

By equation (20·1),

$$K_\nu \parallel \sum_{\lambda=1}^{k-1} a_{\lambda\nu} I_\lambda \qquad (\nu = 0, \dots, k-1) \quad \dots\dots\dots(24\text{·}1).$$

If $|a_{\lambda\rho}| \parallel 0$, then the K's are not linearly independent and hence do not form a fundamental system.

If $|a_{\lambda\rho}| \neq 0$, then we can solve the equations (24·1) for the characteristic invariants, and obtain

$$I_\lambda \parallel \sum_{\nu=0}^{k-1} A_{\lambda\nu} K_\nu \qquad (\lambda = 0, \dots, k-1),$$

where $A_{\lambda\nu}$ is the cofactor of $a_{\lambda\nu}$ in $|a_{\lambda\nu}|$, divided by $|a_{\lambda\nu}|$. It follows that the K's are linearly independent and form a fundamental system.

From this theorem we have an easy method of finding the characteristic invariants, provided that we know already some other fundamental system. We continue with the example of the binary quadratic modulo 2. Dickson* has shown that a fundamental system of residual invariants for this case is given by

$$K_0=1,\quad K_1=b,\quad K_2=abc,\quad K_3=(1-a)(1-b)(1-c).$$

We can tabulate the value of each K for each C as follows:

	C_0	C_1	C_2	C_3
K_0	1	1	1	1
K_1	0	1	0	1
K_2	0	0	0	1
K_3	1	0	0	0

$$|a_{\lambda\rho}| = \begin{vmatrix} 1 & 0 & 0 & 1 \\ 1 & 1 & 0 & 0 \\ 1 & 0 & 0 & 0 \\ 1 & 1 & 1 & 0 \end{vmatrix} \equiv 1;$$

the reciprocal determinant is

$$\begin{vmatrix} 0 & 0 & 0 & 1 \\ 0 & 1 & 1 & 0 \\ 1 & 1 & 0 & 1 \\ 0 & 0 & 1 & 0 \end{vmatrix}.$$

And we have

$$\begin{aligned} I_0 &= K_3=(1+a)(1+b)(1+c), \\ I_1 &= K_2+K_1=b(1+ac), \\ I_2 &= K_0+K_1+K_3=1+b+(1+a)(1+b)(1+c), \\ I_3 &= K_2=abc. \end{aligned}$$

§ 25. Smallest Full Systems.

It is obvious that a fundamental system cannot be a smallest full system on account of the relation (20·4). Indeed any invariant K can be expressed as follows:

$$\begin{aligned} K &= \sum_{j=0}^{m-1} i_j I_j = \sum_{j=1}^{m-1} i_j I_j + i_0\left(1-\sum_{j=1}^{m-1} I_j\right) \\ &= i_0+(i_1-i_0)I_1+\ldots+(i_{m-1}-i_0)I_{m-1}. \end{aligned}$$

* *Madison Colloquium Lectures*, p. 29.

If we are considering the field $GF[p^n]$ whose elements are

$$u_0 = 0,\ u_1,\ u_2,\ \ldots,\ u_{p^n-1},$$

then any invariant can take at most p^n different values, that is, any invariant can characterise at most p^n classes. It follows that s invariants can characterise at most p^{ns} different classes. Suppose that there are k classes; then the minimum number of invariants in a smallest full system will be $t+1$, where t is the greatest power of p^n which is less than k; i.e. $p^{nt} < k \leqslant p^{n(t+1)}$. We shall now show that there always exists a full system with only $t+1$ elements; hence *the number of elements in a smallest full system is* $t+1$.

The method employed in finding the elements of a smallest full system can best be illustrated by studying a simple example. Let us suppose that there are 12 classes $C_0, C_1, \ldots, C_{11}$, and that their characteristic invariants $I_0, I_1, \ldots, I_{11}$ are known. We shall also suppose that the field is $GF[3]$; the elements of this field are 0, 1, 2. Now $3^2 < 11 < 3^3$, hence $t=2$, and there will be three elements D_1, D_2, D_3 in the smallest full system. Let D_1, D_2, D_3 take the following values for the different classes.

	C_0	C_1	C_2	C_3	C_4	C_5	C_6	C_7	C_8	C_9	C_{10}	C_{11}
D_1	0	1	2	0	1	2	0	1	2	0	1	2
D_2	0	0	0	1	1	1	2	2	2	0	0	0
D_3	0	0	0	0	0	0	0	0	0	1	1	1

Then by (20·1)

$$\left.\begin{aligned} D_1 &= I_1 - I_2 + I_4 - I_5 + I_7 - I_8 + I_{10} - I_{11} \\ D_2 &= I_3 + I_4 + I_5 - I_6 - I_7 - I_8 \\ D_3 &= I_9 + I_{10} + I_{11} \end{aligned}\right\} \quad \ldots\ldots(25 \cdot 1).$$

Since the I's are known we can find the D's at once. It is easy to see that the D's form a smallest full system. At first sight it does not seem that equations (25·1) are soluble for the I's, but this is not difficult in view of the relations (20·4) and (20·5) by considering expressions such as D_1^2, D_1D_2, ... E.g., $D_1D_3 = I_{10} - I_{11}$, hence $(D_1D_3)^2 = I_{10} + I_{11}$ so that $D_1D_3 + (D_1D_3)^2 = -I_{10}$.

As an example we consider the binary quadratic mod 2. There will be two invariants in a smallest full system, since from § 19 there are four classes. Give D_1 and D_2 the following values:

	C_0	C_1	C_2	C_3
D_1	0	1	0	1
D_2	0	0	1	1

then, from § 24,

$$D_1 = I_1 + I_3 = b(1 + ac) + abc = b,$$

$$D_2 = I_2 + I_3 = 1 + b + (1 + a)(1 + b)(1 + c) + abc = a + c + ab + ac + bc,$$

giving the same full system as that obtained in § 22. Further

$$D_1 D_2 = I_3, \quad D_1 D_2 + D_1 = I_1, \quad D_1 D_2 + D_2 = I_2, \quad 1 + D_1 + D_2 + D_1 D_2 = I_0.$$

The method illustrated in the above examples is perfectly general.

§ 26. Residual Invariants of Linear Forms.

The following treatment of linear forms is due to Dickson*. Let

$$l_i = a_{i1} x_1 + a_{i2} x_2 + \dots + a_{im} x_m$$

and put

$$A_i = \prod_r (1 - a_{ir}^q) \quad \dots\dots\dots\dots\dots\dots (26{\cdot}1).$$

There are only two classes for a single form l_1, C_0 containing the null form and C_1 containing every other form. We have readily

$$I_0 = A_1, \qquad I_1 = 1 - A_1;$$

also A_1 is seen to form a smallest full system.

We must now consider the classes of a pair of forms l_1 and l_2. We have the class C_0 if $l_1 \equiv l_2 \equiv 0$, C_1 has $l_1 \equiv 0$, $l_2 \not\equiv 0$ and in this case we choose as a type 0, x_1. If l_1 and l_2 are linearly independent, then we have the class C_2 and choose as our type x_1, dx_2, where $d = 1$ if $m > 2$ and $d = D_{12}$, the discriminant of the forms if $m = 2$. We have also the classes $C_{3(c)}$ where l_2 is cl_1 and as type we take x_1, cx_1. For the case where $m = 2$ instead of one class C_2 we have q classes $C_{2(d)}, \dots$. In general let k be a non-zero element of the $GF[p^n]$ and write $l_{ij} = l_i - kl_j$. Using $k^{p^n} \parallel k$ we have

$$A_{ij} = \prod_r [1 - (a_{ir} - ka_{jr})^q] \parallel A_i + A_i(A_j - 1)k^q + \sum_{t=1}^{q} S_{ijt} k^t$$

and the coefficients of powers of k must be invariants. In particular write $V_{ij} = S_{ij1}$.

If $m > 2$, the values of the different invariants for the different classes are as follows:

Class	Types		Invariants			
	l_1	l_2	A_1	A_2	A_1A_2	V_{12}
C_0	0	0	1	1	1	0
C_1	0	x_1	1	0	0	0
C_2	x_1	dx_2	0	0	0	0
$C_{3(c)}$	x_1	cx_1	0	$1 - c^q$	0	c

* *Proc. Lond. Math. Soc.* (2), vol. 7, p. 430 (1909).

From this table we have readily

$$I_0 = A_1 A_2, \qquad I_1 = A_1(1 - A_2), \qquad I_{3(0)} = A_2(1 - A_1),$$

$$I_{3(c)} = 1 - (c - V_{12})^q \quad (c \neq 0), \qquad I_2 = 1 - I_0 - I_1 - \sum_{c/p^n} I_{3(c)}.$$

For the case $m = 2$ the value of $I_{2(d)}$ is

$$1 - (d - D_{12})^q.$$

A smallest full system is given by

$$A_1,\ A_2,\ V_{12}, \quad \text{if} \quad m > 2; \quad A_1, A_2, V_{12}, D_{12}, \quad \text{if} \quad m = 2.$$

Smallest full systems can easily be obtained from the above characteristic invariants. Dickson gives the following fundamental systems of residual invariants of a pair of linear forms l_1 and l_2:

$$1,\ A_1,\ A_2,\ A_1 A_2,\ V_{12}{}^t \qquad (t = 1, 2, \ldots, q) \qquad m > 2,$$

$$A_1,\ A_2,\ A_1 A_2,\ V_{12}{}^t,\ D_{12}{}^t \qquad (t = 1, 2, \ldots, q) \qquad m = 2.$$

We find the classes of a system of λ binary linear forms as follows: We say that $E_{\rho\sigma\tau\ldots}$ follows $E_{rst\ldots}$ if $r < \rho$ or if $r = \rho$, $s < \sigma$ or if $r = \rho$, $s = \sigma$, $t < \tau$, ... First, let not every D_{ij} vanish. Let D_{rs} be the first which does not vanish, then

$$D_{ij} = 0 \quad (i < r), \qquad D_{rk} = 0 \quad (r < k < s), \qquad D_{rs} \neq 0,$$

and by an obvious transformation

$$l_r = x_1, \qquad l_s = cx_2, \qquad c = D_{rs}.$$

Since $D_{ir} = D_{is} = 0$ $(i < r)$, l_i is free of x_1 and x_2.

Since $D_{rk} = 0$ $(r < k < s)$, l_k is free of x_2; therefore we can take as type A_{rs}^{cde}:

$$l_i = 0, \qquad l_r = x_1, \qquad l_k = c_k x_1, \qquad l_s = cx_2, \qquad l_t = -d_t x_1 + e_t x_2$$

$$(i < r < k < s < t \leqslant \lambda; \quad c \neq 0) \qquad \ldots\ldots(26\cdot2).$$

Secondly, if every D_{ij} vanish but not every l, then we can take as type B_f^m:

$$l_i = 0, \qquad l_f = x_1, \qquad l_j = m_j x_1 \qquad (i < f < j \leqslant \lambda) \ \ldots(26\cdot3).$$

Lastly we have the class C_0 containing only null forms ......(26·4).

By giving c, the c_k's, the d_t's, the e_t's and the m_j's, all the values possible, we can in the above manner find all the different classes. Now the products

$$\prod_{k=r+1,\ldots,s-1;\ t=s+1,\ldots,\lambda} D_{rs}^{\gamma} D_{ks}^{\gamma_k} D_{rt}^{\epsilon_t} D_{st}^{\delta_t} \qquad \begin{pmatrix} \gamma = 1, 2, \ldots, q \\ \gamma_k,\ \epsilon_t,\ \delta_t = 0, 1, \ldots, q \end{pmatrix}$$

$$\ldots\ldots(26\cdot5)$$

are linearly independent in the field; also the number of these invariants is equal to the number of classes A_{rs}^{cde}. Similarly

$$1 - A_f, \quad \prod_{j=f+1}^{\lambda} V_{fj}^{\mu_j} \quad (\mu_j = 0, 1, \dots, q : \text{not every } \mu_j = 0)$$
$$\dots\dots(26\cdot6)$$

are linearly independent and of the same number as the classes of type B_f^m. We associate the invariant 1 with the class C_0. The products $(26\cdot5)$ vanish for the classes of B and C_0, also for those classes of A which follow A_{rs}; $(26\cdot6)$ vanish for the classes C_0 and those of B which follow B_f.

A fundamental system of residual invariants of λ binary linear forms is given by

$$1, \quad (26\cdot6), \quad (26\cdot5).$$

By an extension of this method Dickson gives a fundamental system of $\lambda \geqslant m$ linear forms, but for this we would refer the reader to the paper itself. He also treats the case when $\lambda < m$.

§ 27. Residual Invariants of Quadratic Forms.

Our first task in the finding of residual invariants of quadratic forms is to separate the general quadratic form into classes. We see at once that this is a much more difficult task than it was in the binary case. We shall find an essential difference between the cases $p = 2$ and $p > 2$. We shall follow Dickson's methods[*].

First let $p > 2$, then we can write the general quadratic form as

$$q_m = \sum_{i,j=1}^{m} b_{ij} x_i x_j \quad (b_{ij} = b_{ji}), \quad \dots\dots(27\cdot1)$$

and we can choose $b_{11} \neq 0$ with perfect generality. Let the determinant $B = |b_{ij}|$ be of modular rank r, i.e. every minor of order exceeding r is congruent to zero but not every minor of order r is congruent to zero. Now let $b_{11} \mathfrak{b}_i \mathbin{|||} b_{1i}$, then under a transformation of matrix

$$\begin{bmatrix} 1, & -\mathfrak{b}_2, & -\mathfrak{b}_3, & \dots & -\mathfrak{b}_m \\ & 1 & & & \\ & & 1 & & \\ & & & \ddots & \\ & & & & 1 \end{bmatrix} \quad \dots\dots\dots(27\cdot2)$$

we have $q_m = b_{11} x_1^2 + \phi$, where ϕ is independent of x_1. Proceeding in this way we can ultimately replace q_m by

$$a_1 x_1^2 + a_2 x_2^2 + \dots + a_r x_r^2 \quad (a_1 \neq 0, a_2 \neq 0, \dots, a_r \neq 0) \dots(27\cdot3).$$

[*] *Madison Colloquium Lectures*, pp. 4–12 and *Amer. Journ. of Maths.* vol. 30, p. 263 (1908).

Now every a thus obtained can be written congruent to a power of some primitive root ρ of p. After applying a transformation of determinant unity which permutes the $x_1^2, \ldots, x_r^2$ we can assume that $a_1, \ldots, a_s$ are even powers of ρ while $a_{s+1}, \ldots, a_r$ are odd powers of ρ. The transformation

$$\left.\begin{array}{ll} x_i = \rho^k \bar{x}_i & (i < m) \\ x_j = \bar{x}_j, & (j \neq i, j \neq m) \\ x_m = \rho^{-k} \bar{x}_m & \end{array}\right\} \quad \ldots\ldots\ldots\ldots\ldots\ldots (27\cdot4)$$

is of determinant unity. If $r < m$, transformations of the above type replace

$$a_1 x_1^2 + \ldots + a_r x_r^2 \text{ by } x_1^2 + \ldots + x_s^2 + \rho\,(x^2_{s+1} + \ldots + x_r^2) \ldots (27\cdot5).$$

This in turn may be replaced by one of the forms

$$\begin{array}{l} x_1^2 + \ldots + x^2_{r-1} + x_r^2, \\ x_1^2 + \ldots + x^2_{r-1} + \rho x_r^2, \end{array} \quad (0 < r < m) \ \ldots\ldots\ldots\ldots (27\cdot6)$$

by transformations of the types

$$\left.\begin{array}{ll} x_i = \mathfrak{a}_{ii}\bar{x}_i + \mathfrak{a}_{ij}\bar{x}_j & \\ x_j = -\,\mathfrak{a}_{ij}\bar{x}_i + \mathfrak{a}_{ii}\bar{x}_j & \\ x_m = (\mathfrak{a}_{ii}^2 + \mathfrak{a}_{ij}^2)^{-1}\bar{x}_m & \\ x_k = \bar{x}_k & (k \neq i,\ k \neq j,\ k \neq m) \end{array}\right\} \ \ldots\ldots\ldots\ldots (27\cdot7),$$

where $\mathfrak{a}_{ii}$ and $\mathfrak{a}_{ij}$ are chosen so that

$$(\mathfrak{a}_{ii}^2 + \mathfrak{a}_{ij}^2)^{-1} \equiv \rho.$$

If $r = m$, then transformations of type (27·4) replace $\sum_i a_i x_i^2$ by

$$x_1^2 + \ldots + x_s^2 + \rho\,(x^2_{s+1} + \ldots + x^2_{m-1}) + \sigma x_m^2 \quad \ldots\ldots (27\cdot8),$$

where σ is not necessarily equal to ρ. If there be an even number of terms with the factor ρ, transformations of the type (27·7) reduce (27·8) to

$$x_1^2 + x_2^2 + \ldots + x^2_{m-1} + Bx_m^2 \quad \ldots\ldots\ldots\ldots\ldots (27\cdot9)$$

or, if an odd number, to

$$x_1^2 + \ldots + x^2_{m-2} + \rho x^2_{m-1} + \rho^{-1} Bx_m^2 \ \ldots\ldots\ldots\ldots (27\cdot10).$$

(27·10) can be transformed into (27·9) by the substitution

$$\begin{array}{ll} x_{m-1} = -\,\rho^l \bar{x}_m, & \\ x_m = \rho^{-l}\bar{x}_{m-1}, & \\ x_i = \bar{x}_i & (i \neq m,\ i \neq m-1), \end{array}$$

if $B \equiv \rho^{2l+1}$,

or by

$$\begin{array}{ll} x_{m-1} = \mathfrak{a}_{m-1,\,m-1}\bar{x}_{m-1} - \mathfrak{a}_{m,\,m-1}\rho^{2l-1}\bar{x}_m, & \\ x_m = \mathfrak{a}_{m,\,m-1}\bar{x}_{m-1} + \mathfrak{a}_{m-1,\,m-1}\rho\bar{x}_m, & \\ x_i = \bar{x}_i & (i \neq m,\ i \neq m-1), \end{array}$$

where $\mathfrak{a}_{m-1,m-1}$ and $\mathfrak{a}_{m,m-1}$ are chosen such that

$$(\mathfrak{a}^2_{m-1,m-1} + \rho^{2l-2}\mathfrak{a}^2_{m,m-1})^{-1} \mid\mid\mid \rho$$

in the case where $B \mid\mid\mid \rho^{2l}$.

Thus every quadratic form $q_m \bmod p\ (p > 2)$ can be reduced to one of the forms of (27·6) or (27·9) by transformations with determinant unity. For the proof that none of these forms are equivalent we refer the reader to Dickson, *Madison Colloquium Lectures*, page 8.

We cannot of course hope to classify every possible modular form of the m-ary quadratic as we did in the example of § 19. In the more complicated cases we must be content to find a specimen of each class, and we can find whether or not a system of invariants characterise the classes by examining the values which they take for the specimens of each class.

The invariants B and r do not distinguish between the two classes of (27·6) and they therefore do not characterise the classes.

It is known* that a symmetrical determinant of rank $r\ (r > 0)$ has a non-vanishing principal minor of order r. By a suitable transformation we can take this minor as

$$M = \begin{vmatrix} b_{11} & \dots & b_{1r} \\ \dots & \dots & \dots \\ b_{r1} & \dots & b_{rr} \end{vmatrix} \not\equiv 0.$$

It is possible to replace q_m by q_r a function of r variables, and proceeding as before with only r variables we can reduce q_r to

$$x_1^2 + \dots + x^2_{r-1} + Mx_r^2.$$

Now, let M be congruent to ρ^{2l} or ρ^{2l+1}, then the transformation

$$\begin{aligned} x_r &= \rho^{-l}\bar{x}_r \\ x_m &= \rho^{l}\bar{x}_m \qquad (j \neq r,\ j \neq m) \\ x_j &= \bar{x}_j \end{aligned}$$

gives us the two forms of (27·6) and $M^{\frac{q-1}{2}} = \rho^{l(q-1)}$ or $\rho^{\frac{(2l+1)(q-1)}{2}}$ in the respective cases. But these are congruent to $+1$ and -1 respectively. $M^{\frac{q-1}{2}}$ distinguishes therefore between the two forms of (27·6) and we shall call the non-equivalent classes $C_{r,1}$ and $C_{r,-1}$ respectively.

Let us now consider the functions

$$A_r = \{M_1^{\frac{q}{2}} + M_2^{\frac{q}{2}}(1 - M_1^q) + \dots \\ + M_k^{\frac{q}{2}}(1 - M_1^q)\dots(1 - M_{k-1}^q)\} \prod_i (1 - N_i^q),$$

* Dickson, *Annals of Maths.* Ser. 2, vol. 15, pp. 27–28 (1913–14).

where $M_1, \dots, M_k$ denote the principal minors of order r taken in any order, and the N's are the principal minors of order greater than r.

It is easily seen that A_r has the value $+1$ for any form belonging to the class $C_{r,1}$; the value -1 for any form belonging to the class $C_{r,-1}$; and the value 0 for any form not contained in $C_{r,1}$ or $C_{r,-1}$: thus the following invariants characterise the classes

$$A_1, A_2, \dots, A_{m-1}, B$$

as is seen from the following table which illustrates the case $n = 1, p = 3, m = 3$:

Class	Specimen	A_1	A_2	B
C_0	0	0	0	0
$C_{1,1}$	x_1^2	1	0	0
$C_{1,-1}$	$2x_1^2$	-1	0	0
$C_{2,1}$	$x_1^2 + x_2^2$	0	1	0
$C_{2,-1}$	$x_1^2 + 2x_2^2$	0	-1	0
$C_{B=1}$	$x_1^2 + x_2^2 + x_3^2$	0	0	1
$C_{B=2}$	$x_1^2 + x_2^2 + 2x_3^2$	0	0	2

Thus since all the transformations employed have been those with determinant unity, we have as a full system of invariants of the ternary quadratic mod 3,

$$A_1, A_2, B.$$

We notice however that certain of the classes $C_{B=i}$ are equivalent under transformations whose determinant is not congruent to unity, so that a full system of relative invariants is given by

$$A_1, A_2, \dots, A_m,$$

and the non-equivalent classes are

$$C_0, C_{1,1}, C_{1,-1}, C_{2,1}, C_{2,-1}, \dots, C_{m,1}, C_{m,-1}.$$

The treatment of the quadratic where $p = 2$ is essentially different from the case where $p > 2$. For the separation of the classes we refer the reader to Dickson's papers. He shows that an m-ary quadratic is either reducible to a quadratic in less than m variables, or else it is reducible to

$$x_1x_2 + x_3x_4 + \dots + x_{m-2}x_{m-1} + x_m^2,$$

if m be odd, or to one of the two forms

$$x_1x_2 + x_3x_4 + \dots + x_{m-1}x_m,$$

$$x_1x_2 + \dots + x_{m-1}x_m + x_1^2 + \delta x_2^2,$$

provided that $\delta + \delta^2 + \delta^4 + \dots + \delta^{2^{n-1}} \not\equiv 1$, if m be even.

For small values of m we can of course proceed as in § 19.

§ 28. Cubic and Higher Forms.

The quadratic case is rather difficult when compared with the linear; and we may expect that the separation into classes of the cubic, quartic and higher forms will present great difficulty. This can only be accomplished without enormous labour in a few simple cases. Thus while we may find by different methods a number of invariants, it is not possible in general to say whether a system is full or not. The separation into classes is really a subject in itself and we would refer the reader who is interested to a list of papers on the subject which Dickson gives in the last chapter of his *History of the Theory of Numbers.*

§ 29. Relative unimportance of Residual Covariants.

Nothing has been said so far about finding full systems of residual covariants. Indeed there is no method known which can be applied to a general case. Neither of the two methods which can be used for invariants hold when applied to covariants. We notice however that classes of modular forms are completely characterised by rational integral invariants and therefore an invariantive property of a system of modular forms can be expressed by the vanishing of an invariant. This is in contrast to the algebraic case where a property may be expressed by the vanishing of a covariant. The difficulty of finding a means of forming a full system of residual covariants may be linked up with the fact that residual invariants in themselves completely characterise the classes of systems of forms.

§ 30. Non-formal Residual Covariants.

The subject of non-formal residual covariants has not yet been studied. In this type the α's are arbitrary but the a's are elements of the $GF[p^n]$. It is seen therefore that these bear the same relation to the residual covariants as the congruent covariants bear to the formal covariants. This gives us a clue to the study of non-formal concomitants. They are obtained from congruent concomitants by reducing the coefficients of the forms by Fermat's Theorem. If it be true that every congruent covariant is congruent to an algebraic covariant, then it is also true that every non-formal residual covariant can be obtained from an algebraic covariant by reductions of the two types II and III. Also every non-formal residual covariant is a residual covariant, just as every congruent covariant is a formal covariant. It is seen then that these non-formal covariants are not of any special interest, but are merely those residual covariants which are obtained from congruent covariants through reductions by Fermat's Theorem.

PART II

§ 31. Rings and Fields.

A **RING** is a collection of elements $a, b, c, \ldots$, such that for every pair of elements a, b, a sum $a+b$ and a product $a \times b$ are defined, where $a+b$ and $a \times b$ both belong to the ring, and such that the following laws are satisfied:

I. (i) $$a+(b+c)=(a+b)+c,$$

(ii) $$a+b=b+a,$$

(iii) $$a+x=b,$$

has always a solution for x in the ring.

II. $$a \times bc = ab \times c.$$

III. $$a(b+c)=ab+ac.$$

We shall only consider **COMMUTATIVE** rings where the following law also holds.

IV. $$ab=ba.$$

A ring together with the laws which the elements of the ring obey is called an **ALGEBRA**. An algebra which has no divisors of zero is called a **DIVISION** algebra. That is, in a division algebra if $ab=0$ and if $a \neq 0$, then must $b=0$.

A **FIELD** is a collection of the elements of a division algebra for which every equation $ax=b$ has one and only one solution for x provided that a is not zero. We shall denote this unique solution by $x=b/a$.

A field is said to be **FINITE** if it contains only a finite number of elements. Every division ring R (i.e. a ring with no divisors of zero) is contained in its **QUOTIENT FIELD**. The elements of this quotient field L are determined by all the elements given by solutions of the following type of equation:

$$ax=b, \qquad \text{where } a \neq 0,$$

provided that the further conditions hold:

(i) $$a/b=c/d \qquad \text{if } ad=bc,$$

(ii) $$ra/a=r \qquad \text{if } r \text{ belongs to the division ring},$$

(iii) $$(a/b)\times(c/d)=ac/bd,$$

(iv) $$(a/b)+(c/d)=(ad+bc)/bd.$$

Every element of R is an element of L and we say that R is contained in L. Every field which contains an element other than the zero element contains the unit element e which is defined by $ex = x$ where x is any element in the field. There can only be one unit element, for, if e_1 and e_2 were two solutions, then would

$$e_1 a = a, \text{ where } a \neq 0; \quad e_2 b = b, \text{ where } b \neq 0.$$

Hence $e_1 ab = e_2 ab$ and $(e_1 - e_2)\, ab = 0$. It follows that $e_1 = e_2$.

If every element of a ring L is also an element of the ring K, then K is said to contain L. We shall use the sign $\leqslant$ to denote "is contained in"; thus, $L \leqslant K$; K is called a SUPER-RING of L and L is called a SUB-RING of K. If M be a super-ring of L and at the same time a sub-ring of K, then M is said to be a MEDIAL-RING between K and L. Replacing the word "ring" by the word "field" in the foregoing definitions, SUPER-FIELD, SUB-FIELD and MEDIAL-FIELD are defined. If L be a sub-ring of K, we write $L \leqslant K$. If $L < K$, $L \neq K$, i.e. if there exists elements of K which are not elements of L, then L is called an ACTUAL sub-ring of K. The term "actual" is used in other cases with a similar meaning; e.g. if L be an actual super-field of K, then $L > K$, $L \neq K$.

Consider now a ring R containing the elements $a_1, a_2, \ldots$, and let x be an arbitrary quantity, then $A = \sum_i a_{r_i} x^i$ is called an R-POLYNOMIAL. If the sum and product of two R-polynomials are defined in the usual manner, then obviously these R-polynomials form a ring. This ring is called a POLYNOMIAL RING of R and is denoted by $R[x]$. By an obvious extension, polynomial rings containing R-polynomials in several variables can be obtained. Thus

$$R[x_1, x_2] = R_1[x_2], \text{ where } R_1 = R[x_1].$$

Since the field K contains the unit element e, it will also contain the elements $e + e = 2e$, $3e$, ..., $\pm\, ne$. There are two possibilities here; either e, $\pm\, 2e$, $\pm\, 3e$, ..., $\pm\, ne$, ... are all different, in which case the field K is said to have the CHARACTERISTIC zero; or else they are not all different. In this case let $me = ne$, hence $(m - n)\, e = 0$. Let p be the smallest value of $m - n$ for which this is true; then the elements $0, e, 2e, \ldots (p-1)\, e$ are all distinct and K is said to be of characteristic p. We show that K is of characteristic 0 or p by writing $K^{(0)}$ or $K^{(p)}$ as the case may be.

The following theorems are easily obtained*:

THEOREM 1. *It is necessary and sufficient for the vanishing of the derivative of a rational function of x, either that the function is a con-*

* Steinitz, *Algebraische Theorie der Körper*, p. 46.

stant provided that the characteristic of field of the coefficients be zero, or that the function be also a rational function of $y = x^p$, *provided that the characteristic be* p.

THEOREM 2. *If* $f(x)$, *a rational function of* x, *has an r-fold linear factor* $(r \geqslant 1)$, *then the derivative* $f'(x)$ *contains this factor at least* $(r-1)$ *times. If the field to which the coefficients of* $f(x)$ *belong has the characteristic* p *and if* r *is divisible by* p, *then* $f'(x)$ *has the factor at least* r *times, otherwise it has the factor exactly* $r-1$ *times.*

THEOREM 3. *If* $f(x)$ *be a polynomial in* x^{p^f} *and the characteristic of the field be* p, *then* $f(x)$ *is the* p^f*-th power of a polynomial* ϕ *in* x *and* vice versa *where the coefficients of* ϕ *are in a super-field* $L^{(p)}$ *of* $K^{(p)}$.

The proofs of these theorems are left to the reader.

§ 32. Expansions.

A super-field K_2 of a field K_1 can be obtained in the following manner. Choose an arbitrary element α which does not belong to K_1, and let the elements of K_1 be $\beta_1, \beta_2, \ldots, \beta_n$. If α is to be a member of K_2, so must $\alpha+\beta_1$, $\alpha+\beta_2$, $2\alpha+\beta_1$, $\alpha\beta_1$, α^2, $\alpha^3\beta_1+\beta_2$, α/β_1, β_n/α, etc. In this way we obtain a number of new elements and each one can be expressed as a quotient of two polynomials in α with coefficients belonging to K_1. The totality of such quotients, including the cases where one or both of the polynomials do not contain α, form a super-field K_2 and we write $K_2 = K_1(\alpha)$. A super-field of K_2 is also a super-field of K_1 and we write $K_3 = K_2(\gamma) = K_1(\alpha, \gamma)$. Similarly $K_1(\alpha_1, \alpha_2, \ldots, \alpha_n)$ is a super-field of K_1. Every super-field can be built up in this manner. As an example we notice that the field of all complex numbers is obtained from the field of all real numbers by the adjoining of a single element $i = \sqrt{-1}$. The above process leads us to use the word EXPANSION as an alternative for super-field. Some writers use the term EXTENSION as a further alternative.

An expansion is said to be SIMPLE if it is obtained by the adjoining of a single element which does not belong to K_1. It is easily shown that every element of the expansion $K(x_1, \ldots, x_m)$ can be expressed as the quotient of two elements belonging to the polynomial ring $K[x_1, \ldots, x_m]$. Hence

$$K < K[x] \leqslant K(x).$$

There are two principal kinds of expansions. Consider two K-polynomials $A = \sum_i a_i x^i$ and $B = \sum_i b_i x^i$. We shall suppose that $A = B$ but that $A \not\equiv B$; i.e. for some values of i, $a_i \neq b_i$. Since $A = B$,

$A-B=0$, where $A-B$ is a polynomial in x which does not identically vanish. If such a polynomial exist, we say that $K(x)$ is an **ALGEBRAIC** expansion of K, and x is said to be an algebraic element with respect to K. Every algebraic element satisfies an equation with coefficients from the field.

Let $\phi(x)=0$ be the equation of lowest degree which is satisfied by x, or

$$\phi(x)=c_0x^n+c_1x^{n-1}+\ldots+c_n=0 \qquad (c_0 \neq 0,\ n \geqslant 1).$$

Now $\phi(x)$ must be irreducible in the field, for otherwise divisors of zero would exist. Any element of $K(x)$ can now be expressed as a polynomial in x of degree less than or equal to $n-1$. Indeed any element ξ of $K(x)$ is of the form $f(x)/g(x)$ where $f(x)$ and $g(x)$ are K-polynomials and where $(f, g)=e$, the unit element. If ξ be a multiple of $\phi(x)$, then $\xi=0$ since $\phi(x)=0$. If ξ be not a multiple of $\phi(x)$, then, since $g(x)$ may not be a multiple of $\phi(x)$, it is always possible to find a polynomial $h(x)$ such that $h(x)g(x)$ is equal to unity plus a multiple of $\phi(x)$. Hence $h(x)g(x)=1$ since $\phi(x)=0$ and

$$\xi=\frac{f(x)}{g(x)}=\frac{f(x)\,h(x)}{g(x)\,h(x)}=f(x)\,h(x).$$

We can now reduce the degree of $f(x)\,h(x)$ to $n-1$ or lower by means of the equation $\phi(x)=0$. We shall say that x is of **DEGREE** n with respect to the field K if $\phi(x)$ is of the nth degree in x.

For any n elements of $K(x)$ we have therefore n equations of the type

$$a_0+a_1x+\ldots+a_{n-1}x^{n-1}-a=0,$$

so that on eliminating the $x, x^2, \ldots, x^{n-1}$, we have a vanishing determinant and there is therefore always a linear relation with coefficients from K between any n elements of $K(x)$. We have two such relations for any $n+1$ elements and therefore a homogeneous linear relation is obtained by eliminating the constant terms. Every element of an algebraic expansion of K is algebraic with respect to K. This gives us another definition of an algebraic expansion, namely, an expansion L of K is said to be algebraic if every element of L is algebraic with respect to K. In some cases however every $\phi(x)$ is reducible, e.g. if K is the field of all complex numbers. In such a case no algebraic expansion is possible.

Suppose now, on the other hand, that there exist no two elements A and B which are equal unless each a_i is identical with the corresponding b_i. In such a case there will be no polynomial in x which is equal to zero; and $K(x)$ is called a TRANSCENDENTAL expansion of K: further, x is said to be a transcendental element with respect to K.

We have therefore the following criterion: $K(x)$ is an algebraic or a transcendental expansion of K according as $e, x, x^2, \ldots$ are linearly dependent or independent with coefficients from K. An infinite system of elements is said to be linearly dependent or independent with respect to a field K according as there exists or does not exist a finite sub-system of these elements which is linearly dependent with respect to K.

THEOREM 1*. *In a transcendental expansion $K(x)$ every element not contained in K is a transcendental element with respect to K.*

If ξ be an element of $K(x)$, then $\xi = f(x)/g(x)$, where $f(x)$ and $g(x)$ belong to $K[x]$, and without loss of generality we can put $(f, g) = e$, the unit element. Now if ξ were algebraic, then there would exist an equation of the sort

$$c_0 + c_1\xi + \ldots + c_r\xi^r = 0 \qquad (c_r \neq 0,\ c_0 \neq 0),$$

where each c_s belongs to K. Substituting for ξ and multiplying throughout by g^r, we would have

$$c_0 g^r + c_1 f g^{r-1} + \ldots + c_r f^r = 0,$$

or

$$-c_0 g^r = f[c_1 g^{r-1} + \ldots + c_r f^{r-1}],$$

thus g^r would be divisible by f, which is in contradiction to $(f, g) = e$. Hence either ξ is transcendental or else f and g are constants, in which case ξ is contained in K. This proves the theorem.

THEOREM 2†. *If n elements $x_1, x_2, \ldots, x_n$ are adjoined to a field K_0 so that the expansions $K_0(x_1) = K_1$, $K_1(x_2) = K_2, \ldots, K_{n-1}(x_n) = K_n$ are obtained, and if every element x_k is transcendental with respect to K_{k-1} $(k = 1, \ldots, n)$, then is every element x_k transcendental with respect to the field which is obtained by adjoining the $n-1$ remaining elements to K.*

We have seen that every element of an expansion is a quotient of two elements of the corresponding polynomial ring. If now x_k be not transcendental with respect to $K_0(x_1, \ldots, x_{k-1}, x_{k+1}, \ldots, x_n)$, then there will exist a relation

$$c_0 + c_1 x_k + c_2 x_k^2 + \ldots + c_m x_k^m = 0 \quad \text{..............(32·1)},$$

where each c_i is a quotient of two elements of

$$K_0[x_1, \ldots, x_{k-1}, x_{k+1}, \ldots, x_n].$$

* Steinitz, *Algebraische Theorie der Körper*, p. 23.

† *loc. cit.*, p. 26.

Hence multiplying throughout by the L.C.M. of the denominators we can consider each c_i as being an element of $K_0[x_1, \ldots, x_{k-1}, x_{k+1}, \ldots, x_n]$. We can therefore rearrange the equation in powers of x_n and thus obtain

$$C_0 + C_1 x_n + \ldots + C_s x_n^s = 0 \quad \ldots\ldots\ldots\ldots\ldots\ldots(32{\cdot}2),$$

where each C_i is from $K_0[x_1, \ldots, x_{n-1}]$ and therefore from $K_0(x_1, \ldots, x_{n-1})$ so that x_n is not transcendental with respect to $K_0(x_1, \ldots, x_{n-1})$. This is contrary to hypothesis, and so x_k must be transcendental with respect to $K_0(x_1, \ldots, x_{k-1}, x_{k+1}, \ldots, x_n)$. If x_n does not occur in equation (32·2), then we rearrange (32·1) in powers of the x with the greatest suffix and we proceed in the same manner as above.

§ 33. Isomorphism.

If there be a $(1-1)$ correspondence between the elements of two rings R and $\bar{R}$ and if the products and sums of R correspond to the corresponding products and sums of $\bar{R}$, then R and $\bar{R}$ are said to be ISOMORPHIC. The correspondence is written $R \simeq \bar{R}$. If a, b and $\bar{a}$, $\bar{b}$ be corresponding pairs of elements from R and $\bar{R}$ respectively, then $a \simeq \bar{a}$ $b \simeq \bar{b}$, $(a+b) \simeq (\bar{a}+\bar{b})$ and $ab \simeq \bar{a}\bar{b}$. Isomorphic fields are defined in the same way.

Let L and $\bar{L}$ be two expansions of the field K; further, let L and $\bar{L}$ be isomorphic; if the elements of K in L correspond to the elements of K in $\bar{L}$, then L and $\bar{L}$ are said to be EQUIVALENT. It is obvious that two simple transcendental expansions of K are equivalent, for $f(x) \simeq f(y)$, hence $K(x) \simeq K(y)$ and the elements of K are self-corresponding.

§ 34. Finite Expansions.

Let L be an expansion of K, and suppose that there exist n elements in L which are linearly independent with respect to K, but that there do not exist $n+1$ elements in L which are linearly independent, then L is said to be of DEGREE n with respect to K. This is written as follows:

$$[L:K] = n; \ [K:K] = 1.$$

The expansion is said to be FINITE if n is finite.

The following theorems are proved by Steinitz*:

THEOREM 1. *Every finite expansion L of a field K is an algebraic expansion.*

Let $[L:K] = n$ and let α be an element of L: then, since any $n+1$

* Steinitz, *Algebraische Theorie der Körper*, p. 34.

elements of L are linearly dependent, there must be a linear relation between e, α, α^2, ..., α^n or

$$a_0 e + a_1 \alpha + a_2 \alpha^2 + \ldots + a_n \alpha^n = 0.$$

Thus every element of L is algebraic with respect to K; and L is an algebraic expansion. The converse of this theorem is not true.

If there exist n elements of L which are linearly independent with respect to K, then, since any $n+1$ elements are linearly related, every element of L can be expressed as a linear function of these n with coefficients from K. The n linearly independent elements are said to form a BASIS of L. Any other n elements $\beta_1, \ldots, \beta_n$ can be written in terms of the basis elements $\alpha_1, \ldots, \alpha_n$ thus:

$$\beta_i = \sum_{k=1}^{n} a_{ik}\alpha_k \qquad (i = 1, 2, \ldots, n).$$

If the determinant $|a_{ik}|$ does not vanish in the field, then $\beta_1, \ldots, \beta_n$ also are linearly independent and form a second basis of L.

THEOREM 2. *Let $K<L<M$. The necessary and sufficient conditions that M be a finite expansion of K are that M be a finite expansion of L and that L be a finite expansion of K.*

The condition is necessary; for, let $[M:K]=\nu$, then $\nu+1$ elements of M are linearly dependent with respect to K, that is, with coefficients from K and therefore from L. M is therefore a finite expansion of L. Further $\nu+1$ elements of L are $\nu+1$ elements of M and are linearly dependent with respect to K, so that L is a finite expansion of K. The condition is also sufficient; for, let $[L:K]=m$ and $[M:L]=n$, where m and n are both finite, and let $\alpha_1, \ldots, \alpha_m$ be a basis of L with respect to K, and $\beta_1, \ldots, \beta_n$ be a basis of M with respect to L. The mn elements $\alpha_i\beta_k$ are linearly independent with respect to K, for otherwise would $\sum_{i,k} c_{ik}\alpha_i\beta_k = 0$ where each c_{ik} belongs to K. In this case would $\sum_{k=1}^{n} \gamma_k\beta_k = 0$ where $\gamma_k = \sum_{i=1}^{m} c_{ik}\alpha_i$, but this is impossible unless each γ be zero since the γ's are from L. If γ_k be zero, then $\sum_{i=1}^{m} c_{ik}\alpha_i = 0$, which is impossible unless every c_{ik} is zero. It is easy to see that every element of M can be expressed as a homogeneous linear function of the $\alpha_i\beta_k$'s, so that the $\alpha_i\beta_k$'s form a finite basis of M. It follows that M is finite and that

$$mn = [M:K] = \nu.$$

THEOREM 3. *If $K(x)$ is a simple algebraic expansion of K, then $K(x)$ is a finite expansion of K of degree d, where d is the degree of x with* *respect to K.* If x is of degree d with respect to K, then every element α of $K(x)$ can be expressed as follows:

$$\alpha = a_0 + a_1 x + \ldots + a_{d-1} x^{d-1} \qquad (a_i \text{ from } K).$$

Hence $e, x, \ldots, x^{d-1}$ form a basis of $K(x)$ and hence $[K(x) : K] = d$.

THEOREM 4. *An expansion L of K is a finite expansion if, and only if, L can be obtained from K by the adjunction of a finite number of algebraic elements.*

This is sufficient; for if $K = K_0$ and $L = K_0(j_1, j_2, \ldots, j_\nu)$, then put

$$K_1 = K_0(j_1),\ K_2 = K_1(j_2),\ \ldots,\ K_k = K_{k-1}(j_k),\ \ldots,\ L = K_\nu = K_{\nu-1}(j_\nu).$$

Hence j_k is algebraic with respect to K_{k-1}: and K_k is a simple algebraic expansion of K_{k-1} and is therefore a finite expansion of K_{k-1}, by Theorem 3. Hence from Theorem 2, L also is a finite expansion of $K_0 = K$.

The condition is also necessary; for let L be a finite expansion of $K = K_0$ and let $[L : K_0] = n$. If j_1 belongs to L but not to K_0, then is $K_0 < K_0(j_1) \leqslant L$. If $K_0(j_1) < L$, then let j_2 belong to L but not to $K_0(j_1) = K_1$. Hence $K_1 < K_1(j_2) \leqslant L$ and so on. From Theorem 2 the degrees $[K_1 : K_0]$, $[K_2 : K_0]$, $[K_3 : K_0]$ are ascending integers less than or equal to n and hence for a certain finite ν we have the condition

$$K_\nu = L = K_0(j_1, j_2, \ldots, j_\nu).$$

THEOREM 5. *Let S be a collection of elements α which are all algebraic with respect to K; then $K(S)$ is also algebraic with respect to K.*

If S is a finite collection, then, from Theorem 4, $K(S)$ is a finite and therefore an algebraic expansion of K. If S is infinite and β is an element from $L = K(S)$, then β is a rational function of a finite number of elements $\alpha_1, \ldots, \alpha_\nu$ from S with coefficients from K. Thus β is contained in $K(\alpha_1, \ldots, \alpha_\nu)$ and is thus algebraic with respect to K. Therefore $L = K(S)$ is algebraic with respect to K from definition.

THEOREM 6. *Let $K \leqslant L \leqslant M$ and let L be algebraic with respect to K and let M be algebraic with respect to L; then M is algebraic with respect to K also.*

For, if β is an element of M of degree n with respect to L, then β satisfies an equation

$$\alpha_0 + \alpha_1 \beta + \ldots + \alpha_n \beta^n = 0 \qquad (\alpha_i \text{ from } L).$$

Now every α_i is algebraic with respect to K, therefore from Theorem 4 $L' = K(\alpha_0, \ldots \alpha_n)$ is a finite expansion of K, and therefore β is algebraic with respect to L'. Thus $L'(\beta)$ is a finite expansion of L' and therefore of K. Therefore from Theorem 1, β is algebraic with respect to K.

§ 35. Transcendental and Algebraic Expansions*.

We shall now make a new definition of a transcendental expansion. Every expansion which is not an algebraic expansion shall be called a transcendental expansion. We find it necessary first to reconsider the expression "algebraic dependency." Let $L > K$ and let S be a system of elements from L, and let a be an element from L. Now S determines an expansion $K(S)$ of K for which $K \leqslant K(S) \leqslant L$. If now a is algebraic with respect to $K(S)$, then we say that a is ALGEBRAICALLY DEPENDENT upon S. Therefore a satisfies an equation with coefficients which are rational functions of the elements of S with coefficients from K. Further, a system T of elements a from L is said to be algebraically dependent upon a system S from L if every element a of T is algebraically dependent on S. In this case T is entirely contained in $K(S)$.

If a is algebraically dependent on S and

$$A_0 + A_1 a + \dots + A_{n-1} a^{n-1} + a^n = 0$$

expresses this dependence, then the elements A_i belong to $K(S)$ and are therefore rational functions of a certain finite number of elements $s_1, \dots, s_\nu$. That is, these coefficients A_i are already contained in $K(s_1, \dots, s_\nu) = K(S')$, where S' is a finite sub-system of S. Hence we have

THEOREM 1. *If a is algebraically dependent on S, then there exists a finite sub-system S' of S such that a is algebraically dependent on S'.*

It is seen from this theorem that when we are considering the notion of one system T being dependent upon another system S, we may always consider T and S to be finite collections of elements.

THEOREM 2. *If S_3 is algebraically dependent on S_2, and S_2 is algebraically dependent on S_1, then S_3 is algebraically dependent upon S_1.*

Let $K(S_1) = K_1$ and $K(S_2) = K_2$. From § 34, Theorem 5, we know that $K_1(S_2)$ is algebraically dependent on K_1 and $K_2(S_3)$ is algebraically dependent on K_2. Now $K_2 < K_1(S_2)$, hence the elements of $K_2(S_3)$ are algebraic with respect to $K_1(S_2)$ and therefore with respect to K_1 by § 34, Theorem 6. Thus the elements of $K_2(S_3)$ are algebraic with respect to $K(S_1)$ and the elements of S_3 are algebraic with respect to $K(S_1)$. It follows that S_3 is algebraically dependent on S_1.

If the system S_1 is algebraically dependent on S_2, and S_2 is algebraically dependent on S_1, then S_1 and S_2 are said to be EQUIVALENT

* Steinitz, *Algebraische Theorie der Körper*, pp. 114–119.

systems. [N.B. The reader must be careful to distinguish between equivalent systems and equivalent expansions.] If S_1 and S_2 are equivalent systems, we write $S_1 \sim S_2$. From Theorem 2, it is seen that if $S_1 \sim S_2$ and if $S_2 \sim S_3$, then $S_1 \sim S_3$.

A system S is said to be **ALGEBRAICALLY REDUCIBLE** with respect to K either if S contains a single element which is algebraic with respect to K, or if S is equivalent to an actual sub-system of itself. If S is not algebraically reducible it is said to be **IRREDUCIBLE**. A reducible system S contains at least one element that is algebraically dependent upon the others. This follows from Theorem 1.

THEOREM 3. *Every sub-system S' of an irreducible system S is irreducible.*

THEOREM 4. *Every reducible system S contains a finite sub-system S' which is reducible.*

A transcendental expansion is said to be **PURE** if it is obtained by the adjoining of an irreducible system. Let $S = x_1, \ldots, x_n$ be a finite irreducible system, then $L = K(x_1, \ldots, x_n)$ is the field of all rational functions of n unknowns. If S be infinite we can show that L is equivalent to $K(x_1, x_2, \ldots)$ where there are an infinity of x's within the bracket. Let $S_1 = x_1, \ldots, x_n$ and $S_2 = y_1, \ldots, y_n$ be two finite irreducible systems with the same number of elements.

THEOREM 5. *$K(S_1)$ is equivalent to $K(S_2)$.* That is, $K(S_1) \simeq K(S_2)$ and every element of K is self-corresponding. We can obtain $K(S_1)$ in the following manner:

$$K_1 = K(x_1),\ K_2 = K_1(x_2),\ \ldots,\ K(S_1) = K_n = K_{n-1}(x_n),$$

and similarly

$$\overline{K}_1 = K(y_1),\ \overline{K}_2 = \overline{K}_1(y_2),\ \ldots,\ K(S_2) = \overline{K}_n = \overline{K}_{n-1}(y_n).$$

It has been shown in § 33 that $K(x_1) \simeq K(y_1)$ and so by an n-fold application, $K(S_1) \simeq K(S_2)$. This theorem may also be proved for the case where S_1 and S_2 are infinite systems.

Suppose now that the system S is irreducible with respect to K and that a is an additional element.

THEOREM 6. *If S is irreducible but $S + a$ is reducible, then a is algebraically dependent upon S.*

From Theorem 4, $S + a$ contains a finite reducible sub-system T and T must contain a. Let $T = T' + a$. Now T' is a finite sub-system of

S and is therefore irreducible. Suppose that $T' = x_1, \ldots, x_n$ and let $K_1 = K(x_1), \ldots, K_n = K_{n-1}(x_n) = K(T')$, where every field K_ν is a pure transcendental expansion of $K_{\nu-1}$. Now, if $K_\nu(a)$ were transcendental with respect to K_ν, then, from § 32, Theorem 2, every element of T would be transcendental with respect to the n remaining elements of T; i.e. T would be irreducible: which is a contradiction. Therefore $K_n(a)$ is algebraic with respect to K_n so that a is algebraic with respect to K_n and therefore with respect to T' and finally with respect to S.

THEOREM 7. *Every system S of elements from $L > K$, of which not every element a is algebraic with respect to K, is equivalent to an irreducible sub-system $S' \leqslant S$.*

We may suppose that S contains no elements which are algebraic with respect to K, for otherwise we can neglect these. Thus every a of L is transcendental with respect to K. If S is a finite system the theorem is obvious. We shall not require this theorem for the case where S is an infinite system, so for the proof of this case we shall refer the reader to Steinitz*.

If $L = K(S)$ is a transcendental expansion of K, then from the theorem just proved $S \sim S'$ where S' is irreducible. Thus every element of S is algebraic with respect to $K(S')$, and $K(S)$ is an algebraic expansion of $K(S')$. Hence follows

THEOREM 8. *Every transcendental expansion is obtained by an algebraic expansion following on a pure transcendental expansion.*

Let $L = K(S)$ where S is equivalent to the irreducible S'. If the system S' has n elements, then n is said to be the TRANSCENDENTAL DEGREE of L with respect to K. If S' contains no elements, i.e. if L is algebraic with respect to K, then we say that the transcendental degree of L with respect to K is zero. If the transcendental degree of L is n, then there exists in L at least one irreducible system of n elements but no irreducible system of more than n elements.

§ 36. Rational Basis Theorem of E. Noether.

THEOREM†. *If $\{f\}$ is a collection of rational functions $f(x_1, \ldots, x_n)$ of n indeterminates $x_1, \ldots, x_n$ with coefficients from a field K, then from $\{f\}$ it is possible to choose a finite number of functions $f_1, \ldots, f_m$ such that every f is a rational function of the $f_1, \ldots, f_m$ with coefficients from the field K.*

* *Loc. cit.*

† *Gött. Nach.* 1926, Heft 1, p. 28.

Such a system of functions $f_1, \ldots, f_m$ is called a **RATIONAL BASIS.** The above theorem is trivial if $\{f\}$ is a finite collection.

I. Now $K(x_1, \ldots, x_n)$ is a pure transcendental expansion of K and is of transcendental degree n. The total of all the rational functions of the f's from $\{f\}$ with coefficients from K also gives a transcendental expansion of K, namely $L = K(\{f\})$. Obviously

$$K < L \leqslant K(x_1, \ldots, x_n).$$

The transcendental degree h of L is then $\geqslant 1$ and $\leqslant n$. Now let $y_1, \ldots, y_h$ be an irreducible system S of L. If it can be shown that L is a finite and therefore an algebraic expansion of $H = K(y_1, \ldots, y_h)$, then the theorem is proved, for if L is a finite expansion of H, then will $L = H(L_1, \ldots, L_l)$ from Theorem 4, § 34 ; i.e. L is obtained by the adjunction of a finite number of elements $L_1, \ldots, L_l$ from L to H. Hence $L = K(y_1, \ldots, y_h, L_1, \ldots, L_l)$. This last shows that every element from L and therefore every f from $\{f\}$ is a rational function of $y_1, \ldots, y_h, L_1, \ldots, L_l$ with coefficients from K, and since the $y_1, \ldots, y_h$ are also elements of L, i.e. are also rational functions of the f's from $\{f\}$, therefore those f's which are used for the y's and the L's form a rational basis for $\{f\}$.

II. It remains therefore to prove that if $K < L \leqslant K(x_1, \ldots, x_n)$ and if L be of transcendental degree h, less than or equal to n with respect to K with an irreducible system $S = y_1, \ldots, y_h$, then L is a finite algebraic expansion of $H = K(y_1, \ldots, y_h)$. If $h = n$ then H is of the same transcendental degree as $K(x_1, \ldots, x_n)$. Since S is an irreducible system of L and therefore of $K(x_1, \ldots, x_n)$ also, hence every element of $K(x_1, \ldots, x_n)$, and thus x_i itself, is now algebraic with respect to H, since every system $y_1, \ldots, y_n, x_i$ is reducible with respect to K. It follows that every element of L and of $K(x_1, \ldots, x_n)$ also is algebraically dependent on H. Hence L and $K(x_1, \ldots, x_n)$ are algebraic expansions of H. Now since

$$K(x_1, \ldots, x_n) = K(y_1, \ldots, y_n, x_1, \ldots, x_n),$$

then $K(x_1, \ldots, x_n)$ is obtained from H by the adjunction of a finite number of algebraic elements $x_1, \ldots, x_n$ and is therefore by Theorem 4, § 34, a finite algebraic expansion of H. But since every element of H is also an element of L, therefore $H \leqslant L \leqslant K(x_1, \ldots, x_n)$.

Hence L also is a finite algebraic expansion of H. The Rational Basis Theorem has now been proved for the case where $h = n$.

Suppose next that $h < n$. S is an irreducible system in L and in $K(x_1, \ldots, x_n)$. The suffixes of the x's can so be chosen that

$$y_1, \ldots, y_h, x_{h+1}, \ldots, x_n = S + X$$

is an irreducible system of $K(x_1, \ldots, x_n)$, for otherwise would the Jacobian $\dfrac{\partial(y_1, \cdots, y_h, x_{i_{h+1}}, \cdots, x_{i_n})}{\partial(x_1, \cdots, x_n)}$ vanish for all $i_{h+1}, \cdots, i_n$, so that the y's would not form an irreducible system of $K(x_1, \ldots, x_n)$. The system $X = x_{h+1}, \cdots, x_n$ is therefore irreducible with respect to $y_1, \cdots, y_h$ and hence with respect to H. We may suppose then that X is an irreducible system with respect to L also; for otherwise x_{h+1}, for example, would be algebraically dependent upon L, i.e. on certain elements from L, so that x_{h+1} would be algebraic with respect to $y_1, \ldots, y_h$, since every element of L is algebraic with respect to S. Hence X is algebraically independent or irreducible with respect to S, H and L, and hence with respect to every medial field M between H and L, where $H \leqslant M \leqslant L$. It follows that

$$\overline{M} = M(x_{h+1}, \cdots, x_n),\ \overline{L} = L(x_{h+1}, \cdots, x_n) \text{ and } \overline{K} = K(x_{h+1}, \cdots, x_n)$$

are pure transcendental expansions of M, L and K respectively and S is therefore irreducible with respect to $\overline{K}$: for if this were not the case then we should have an algebraic equation with coefficients from $\overline{K}$ relating the $y_1, \ldots, y_h$; and $y_1, \cdots, y_h, x_{h+1}, \cdots, x_n$ would not then be an irreducible system of $K(x_1, \ldots, x_n)$. We can now choose $\overline{K}$ in place of K as the coefficient field. Since $K < L < K(x_1, \ldots, x_n)$, therefore $\overline{K} < \overline{L} < K(x_1, \ldots, x_n)$. But the transcendental degree of $\overline{L}$ with respect to $\overline{K}$ is h and that of $\overline{K}(x_1, \ldots, x_h)$ with respect to $\overline{K}$ is also h. Therefore $\overline{L}$ and $\overline{K}(x_1, \ldots, x_h)$ or $\overline{L}$ and $K(x_1, \ldots, x_n)$ are of the same transcendental degree h with respect to $\overline{K}$. But the theorem has been proved for the case where $h = n$, so that $\overline{L}$ is a finite algebraic expansion of $\overline{K}$ with a rational basis $y_1, \cdots, y_h, z_1, \cdots, z_k$, say. Thus

$$\overline{L} = \overline{K}(y_1, \cdots, y_h, z_1, \cdots, z_k) \quad \text{..............(36·1).}$$

We now consider the field

$$M = K(y_1, \cdots, y_h, z_1, \cdots, z_k) \quad \text{..............(36·2).}$$

We may suppose here that $y_1, \cdots, y_h, z_1, \cdots, z_k$ are from L. $y_1, \cdots, y_h$ certainly are, also $\overline{L}$ is a finite algebraic expansion of $\overline{K}$ and therefore $h + k$ elements are adjoined to $\overline{K}$ which do not belong to $\overline{K}$. They can therefore be chosen belonging to L. Obviously

$$H < M \leqslant L \quad \text{......................(36·3).}$$

Formulae (36·1) and (36·2) give $\overline{M} = \overline{L}$, so that L is contained in $\overline{M}$ for $L < \overline{L}$, thus $L < \overline{M}$. It follows from (36·3) that

$$M \leqslant L < \overline{M} \quad \text{......................(36·4).}$$

Now every element of $\overline{M}$, which is not contained in M already, is transcendental with respect to M, since $\overline{M}$ is a pure transcendental expansion of M. Every element of L which does not belong to M is transcendental with respect to M. Also every element of L is algebraic with respect to $y_1, \ldots, y_h$ and therefore with respect to M, so that $M = L = K(y_1, \ldots, y_h, z_1, \ldots, z_k)$, and L is a finite algebraic expansion of H. The theorem of E. Noether is therefore proved for all cases.

We have also the corollary*. *If* $K < L < M < K(x_1, \ldots, x_n)$ *and* M *is an algebraic expansion of* L, *then must* M *be a finite expansion of* L, for M has a finite rational basis by the last theorem.

§ 37. The Fields $K^{p \pm f}$.

The operation of finding the pth root of an element of a $K^{(p)}$ is unique, for if there were two pth roots of α, let them be a and b. Hence $\alpha = a^p = b^p$ and therefore $0 = a^p - b^p = (a-b)^p$, whence $a = b$. We can write the pth root of α uniquely as $\alpha^{\frac{1}{p}}$. If α belongs to K it may or may not happen that $\alpha^{\frac{1}{p}}$ also belongs to K. We shall consider a case where this does not happen.

Let $L = K(\mu)$ be a simple transcendental expansion of a field K with characteristic p. L also must be of characteristic p. Let α be an element of L but not of K and of grade h [the grade of α is maximum (degree of f; degree of g) where $\alpha = f(\mu)/g(\mu)$]. It can be shown that α^p is of grade hp, which is certainly greater than unity. The equation $\alpha^p = \mu$ would then reduce to an equation in μ with coefficients from K so that μ would be algebraic with respect to K. This is impossible, hence the equation $\alpha = \mu^{\frac{1}{p}}$ is impossible, that is, L does not contain $\mu^{\frac{1}{p}}$.

A field K is said to be PERFECT† if every polynomial $F(x)$ with coefficients in the field and with repeated linear factors is reducible in the field. A field K is said to be IMPERFECT if there exist in K irreducible polynomials with repeated zeros in a super-field of K.

Obviously every field of characteristic zero is perfect. For fields of characteristic p we have the following theorem:

There are perfect and imperfect $K^{(p)}$*'s. A* $K^{(p)}$ *is perfect if* $a^{\frac{1}{p}}$ *belongs to* K *provided that* a *belongs to* K. *In every other case* $K^{(p)}$ *is imperfect.*

Suppose then that K is a $K^{(p)}$ and that γ is from K, then from Theorem 3, § 31,

$$x^{p^f} - \gamma = (x - \delta)^{p^f},$$

* *Gött. Nach., loc. cit.* p. 31.

† Steinitz, *loc. cit.* p. 50.

where δ is either from K or from an algebraic expansion of K. We now try to find irreducible factors ϕ of $x^{p^f}-\gamma$. Suppose then that $\phi=(x-\delta)^{p^r}=x^{p^r}-\delta^{p^r}$ $(0\leqslant r\leqslant f)$ be an irreducible factor. We can only have $r=0$ if $\delta=\gamma^{p^{-f}}$ belongs to K. Otherwise $r>0$, in which case we have an irreducible polynomial with p^r repeated roots, and hence $K^{(p)}$ is imperfect.

Suppose now that both γ and $\gamma^{\frac{1}{p}}$ belong to the $K^{(p)}$ and that $f(x)$ is a polynomial which has coefficients from the $K^{(p)}$ and which has repeated zeros. Then is (f, f') of at least the first degree in x. If $f'(x)\neq 0$, then the degree of (f, f') is less than the degree of $f(x)$, and $f(x)$ is therefore reducible. If $f'(x)=0$, then

$$f(x)=a_0+a_1x^p+\ldots+a_\nu x^{\nu p} \qquad (\nu p=n).$$

But by supposition $a_0^{\frac{1}{p}}, a_1^{\frac{1}{p}}, \ldots$ are all elements of $K^{(p)}$, so that

$$f(x)=(a_0^{\frac{1}{p}}+a_1^{\frac{1}{p}}x+\ldots+a_\nu^{\frac{1}{p}}x^\nu)^p$$

and is therefore reducible: hence $K^{(p)}$ in this case is a perfect field.

This leads us to consider the fields which we shall designate by $K^{p^{\pm f}}$. We suppose that K has the characteristic p, then

$$a^p\pm b^p=(a\pm b)^p;\ a^pb^p=(ab)^p;\ a^p:b^p=(a:b)^p.$$

We see therefore that the pth powers of the elements of K constitute a sub-field of K which we shall denote by K^p. (N.B. $K^{(p)}$ is not the same as K^p.) K and K^p are seen to be isomorphic, or $K\simeq K^p$. It is obvious also that $K^p\leqslant K$.

If we now add to K every element $a^{\frac{1}{p}}$ as a new element whenever a is an element of K but not of K^p, we obtain a field L such that $K=L^p$. For this reason we write $L=K^{\frac{1}{p}}$. We can repeat either of the above processes as often as we wish and obtain the fields $K^{p^{\pm f}}$. Such fields as $K^{p^{-f}}$ are expansions of K and we shall call them RADICAL fields of K. In the same way we can obtain radical rings. If L be a radical field of K, then there must be a minimum exponent m such that the p^m-th power of every element in L is an element of K. In this case m is called the EXPONENT of L. Similarly we define the exponent f of an element a of L. It is obvious that $f\leqslant m$.

§ 38. Expansions of the First and Second Sorts.

We can divide algebraic expansions into two sorts*. Let $g(x)$ be a polynomial which is irreducible in the field K. We say that $g(x)$ is of the FIRST SORT if it has only simple zeros when considered in an expan-

* Steinitz, *loc. cit.* p. 62.

sion of K. Now let the element γ be algebraic with respect to K; then γ satisfies an equation of the type $g(x)=0$ which is irreducible in K. If $g(x)$ is of the first sort, then we say that γ is an **ALGEBRAIC ELEMENT OF THE FIRST SORT** with respect to K. An algebraic expansion L of K is said to be of the first sort, if every element of L is an algebraic element of the first sort with respect to K.

We see that every algebraic element with respect to a perfect field is of the first sort. Suppose then that K is a $K^{(p)}$ and is imperfect. Suppose also that $g(x)$ is a polynomial which is irreducible in K. We have seen that there is associated with an element α from the radical fields $K^{\frac{1}{p}}$, $K^{\frac{1}{p^2}}, \ldots$ an exponent f, such that α^{p^f} lies in K but $\alpha^{p^{f-1}}$ does not. We extend this notion to the case where α does not actually belong to a radical field, as follows:

Either $g(x)=G(x^{p^f})$, $f>0$, i.e. $g(x)$ is a polynomial in x^{p^f} for some maximum f; or $f=0$ and $g'(x)$ is not identically zero and has thus no common factor with $g(x)$, so that $g(x)=0$ has only simple zeros and is therefore of the first sort. If $f>0$, $g(x)$ is the p^f-th power of a polynomial in $K^{\frac{1}{p^f}}$ and is not of the first sort. Let this polynomial be $h(x)$ In this case f is called the exponent of $h(x)$, for the p^f-th power of $h(x)$ belongs to K while the p^{f-1}-th power does not.

An algebraic expansion of K is said to be of the **SECOND SORT** if it is not of the first sort.

THEOREM. *If L is an algebraic expansion of an imperfect field K, then there exists a medial field L_0, i.e. $K \leqslant L_0 \leqslant L$, such that L_0 is of the first sort with respect to K and such that L is a radical field of L_0. Further L_0 is the aggregate of all the elements of L which are of the first sort with respect to K.*

This theorem holds for infinite algebraic expansions, but it is only proved here for the finite case. Let us call L_0 the system of all elements in L which are of the first sort with respect to K; then obviously $K(L_0)=K(L_0')$, where L_0' is obtained by removing from L_0 every element which is also contained in K. Further, by Abel's Theorem* on primitive elements, a single element γ can be found such that $K(L_0')=K(\gamma)$, where γ is algebraic and of the first sort with respect to K. Let α be any element of $K(\gamma)$ and let

$$[K(\alpha):K]=m_1, \qquad [K(\gamma):K(\alpha)]=m_2,$$
$$[K(\alpha^p):K]=n_1, \qquad [K(\gamma):K(\alpha^p)]=n_2.$$

* Steinitz, *loc. cit.* p. 52.

Hence $m_1 m_2 = n_1 n_2 = [K(\gamma) : K]$ and since $K(\alpha^p)$ is a sub-field of $K(\alpha)$, therefore

$$m_1 \geqslant n_1, \qquad m_2 \leqslant n_2.$$

Now let $$x^{m_2} + b_1 x^{m_2-1} + \ldots + b_{m_2}$$

be the irreducible function which vanishes when $x = \gamma$, then will

$$x^{m_2} + b_1^p x^{m_2-1} + \ldots + b_{m_2}^p$$

vanish for $x = \gamma^p$ and the coefficients $b_1^p, \ldots, b_{m_2}^p$ belong to the field $K(\alpha^p)$; therefore

$$n_2 = [K(\gamma) : K(\alpha^p)] = [K(\gamma^p) : K(\alpha^p)] \leqslant m_2.$$

Hence $m_2 = n_2$ and $m_1 = n_1$, that is, α and α^p are of the same degree with respect to K. Since this is so, $[K(\alpha); K(\alpha^p)] = 0$. But, by a theorem of Steinitz*, $[K(\alpha); K(\alpha^p)] = p$ if the exponent of α is greater than zero. It follows that the exponent is zero and so α is of the first sort. Hence $K(\gamma)$ is an expansion of the first sort. It has been proved that $K(L_0)$ contains elements of the first sort only, hence $K(L_0) = L_0$ and L_0 is a medial field between K and L or

$$K \leqslant L_0 \leqslant L.$$

Now if δ be an element of L of exponent f with respect to K, then $g(x) = G(x^{p^f}) = G(y)$, say, and $G(y)$ is certainly irreducible in K. Thus $G(y)$ has only simple roots, so that $y = x^{p^f}$ is of the first sort and therefore an element of L_0, i.e. every element of L is a radical element of L_0 and therefore L is a radical field of L_0. It is easy to show that L_0 is uniquely determined by L.

If L is a finite algebraic expansion of an imperfect field K, then is $L = K(\delta_1, \ldots, \delta_m)$. Suppose f_i is the exponent of δ_i and that the maximum f_i is f, then f is the exponent of the expansion of L and $L \leqslant L_0^{\frac{1}{p^f}}$.

§ 39. The Theorem on Divisor Chains†.

Let R be a commutative ring with elements $a, b, c, \ldots$, with a unit element and with no divisors of zero. We shall consider in particular sub-rings $\mathfrak{a}$ of R which have the following properties: (1) if a and b belong to $\mathfrak{a}$, so do $a + b$ and $a - b$; (2) if a belongs to $\mathfrak{a}$ and r belongs to R, then ra belongs to $\mathfrak{a}$. Thus $\mathfrak{a}$ is the total of all elements such as $r'a + r''b + r'''c + \ldots$ and is said to be an IDEAL in R. We write

$$\mathfrak{a} = (a, b, c, \ldots).$$

An ideal $\mathfrak{a}$ is said to divide the ideal $\mathfrak{b}$ if every element b of $\mathfrak{b}$ is contained in $\mathfrak{a}$. This is written $\mathfrak{b} \leqslant \mathfrak{a}$. This idea of division is explained

* *Algebraische Theorie der Körper*, p. 63.

† Van der Waerden, *Moderne Algebra*, vol. II, p. 25.

more fully by van der Waerden*. With this definition of division the H.C.F., written $(\mathfrak{a}, \mathfrak{b}, \mathfrak{c}, \ldots)$, and L.C.M., written $[\mathfrak{a}, \mathfrak{b}, \mathfrak{c}, \ldots]$, of the ideals $\mathfrak{a}, \mathfrak{b}, \mathfrak{c}, \ldots$, are obtained. An ideal $\mathfrak{a} = (a_1, a_2, \ldots, a_n)$ is said to be **FINITE** if n is finite. $a_1, a_2, \ldots, a_n$ are said to form a **BASIS** of the ideal $\mathfrak{a}$.

We shall only consider rings in which every ideal is finite, and we shall now prove the following theorem. *Every ideal of a ring has a finite basis if, and only if, there exists no chain of ideals* $\mathfrak{a}_1 < \mathfrak{a}_2 < \ldots$, *where* $\mathfrak{a}_{i+1}$ *is an actual divisor of* $\mathfrak{a}_i$, *which chain does not come to an end after a finite number of steps.*

This is known as the theorem of divisor chains. The proof is as follows. We shall first suppose that there exists no infinite chain of ideals

$$\mathfrak{a}_1 < \mathfrak{a}_2 < \mathfrak{a}_3 < \ldots,$$

where $\mathfrak{a}_{i+1}$ is an actual divisor of $\mathfrak{a}_i$. Now if $\mathfrak{a}$ be an ideal without a finite basis, then let a_1 be an element of $\mathfrak{a}$. Since $\mathfrak{a} \neq (a_1)$, then $\mathfrak{a}$ must contain an element a_2 such that $\mathfrak{a}_2' = (a_1, a_2)$ is an actual divisor of $\mathfrak{a}_1' = (a_1)$. Since $\mathfrak{a} \neq (a_1, a_2)$, there must exist an a_3 such that

$$\mathfrak{a}_3' = (a_1, a_2, a_3)$$

is an actual divisor of $\mathfrak{a}_2'$, and so on. We would thus obtain an infinite chain of ideals in contradiction to hypothesis: we were therefore in error in supposing that an ideal existed without a finite basis.

Suppose now that every ideal has a finite basis and let

$$\mathfrak{a}_1 < \mathfrak{a}_2 < \mathfrak{a}_3 < \ldots$$

be an infinite chain of ideals such that each is an actual divisor of the preceding. Let $\mathfrak{d} = (\mathfrak{a}_1, \mathfrak{a}_2, \ldots)$ be the H.C.F. of all the ideals $\mathfrak{a}_1, \mathfrak{a}_2, \ldots$. Now $\mathfrak{d}$ is itself an ideal; for if a and b are two elements of $\mathfrak{d}$, then suppose that a belongs to $\mathfrak{a}_i$ and that b belongs to $\mathfrak{a}_j$. If N be the greater of i and j, then both a and b must lie in $\mathfrak{a}_N$ and hence $a+b$, $a-b$ and ra belong to $\mathfrak{a}_N$, where r is an element from R. It follows that $a+b$, $a-b$ and ra lie in $\mathfrak{d}$, hence $\mathfrak{d}$ is an ideal and by hypothesis will have a finite basis $d_1, \ldots, d_\rho$. Each d_i must belong to some ideal $\mathfrak{a}_{\nu_i}$. Let the greatest ν_i be M; then every element of the basis belongs to $\mathfrak{a}_M$ and hence $\mathfrak{d}$ belongs to $\mathfrak{a}_M$; that is $\mathfrak{d} \leqslant \mathfrak{a}_M$. But, since $\mathfrak{d}$ is the H.C.F. of all the ideals $\mathfrak{a}_i$, $\mathfrak{d} \geqslant \mathfrak{a}_M$. It follows that $\mathfrak{d} = \mathfrak{a}_M$ and in the same way it may be shown that $\mathfrak{d} = \mathfrak{a}_{M+t}$ where t is a positive integer. Hence

$$\mathfrak{a}_M = \mathfrak{a}_{M+1} = \mathfrak{a}_{M+2} = \ldots,$$

and the divisor chain is not an infinite one, since $\mathfrak{a}_{M+t}$ is not an actual divisor of $\mathfrak{a}_M$. This concludes the proof of the theorem.

* *Loc. cit.* p. 29.

§ 40. R-Modules.

Let S be a super-ring of R, where both S and R are commutative rings with a unit element and with no divisors of zero. Consider systems M of S which contain R and which have the following properties:

1. If α and β belong to M, so do $\alpha + \beta$ and $\alpha - \beta$.

2. If α belongs to M and r to R, then $r\alpha$ belongs to M; thus M is the aggregate of elements $r'\alpha + r''\beta + \ldots$ of S, where $\alpha, \beta, \ldots$ are from S and $r', r'', \ldots$ are from R. A system M of the above sort is called an **R-MODULE.** $\alpha, \beta, \ldots$ form the **BASIS** of M and if the basis has a finite number of elements, then we say that M is a finite R-module.

If every element of a module M is also contained in a module N, then we say that M is divisible by N. If a module M has a basis $\mu_1, \ldots, \mu_m$ and a module N has a basis $\nu_1, \ldots, \nu_n$, then the module with the basis $\mu_1\nu_1, \ldots, \mu_i\nu_j, \ldots, \mu_m\nu_n$ is called the product of M and N and is written MN. Note that if M is divisible by N, then $M \leqslant N$.

Let M be a finite R-module whose basis is $\xi_1, \xi_2, \ldots, \xi_k$ from S. Every R-module A which contains R and is contained by M must also be finite. *The theorem of divisor chains holds for every R-module A provided that it holds for ideals in R**.

Without lack of generality we may suppose that $\xi_1, \ldots, \xi_k$ are linearly independent. We shall say that an element μ of M is "of the length i" if $\mu = r_1\xi_1 + r_2\xi_2 + \ldots + r_i\xi_i$ for some values of the $r_1, \ldots, r_i$, where $r_i \neq 0$ and i is a minimum, and if no such equation holds for $h < i$. There are therefore in M elements of the lengths 1, 2, ..., k, e.g. $\xi_1, \ldots, \xi_k$. Now the system of all the elements in A of length $\leqslant i$ also form an R-module, namely A_i, and every element in A_i is of the form

$$a_{1i}\xi_1 + a_{2i}\xi_2 + \ldots + a_{ii}\xi_i \quad \ldots\ldots\ldots\ldots\ldots\ldots(40{\cdot}1).$$

The system of elements a_{hi}, all from R, form an ideal $\mathfrak{a}_i$ in R. We obtain in this manner a sequence of ideals

$$\mathfrak{a}_1 \leqslant \mathfrak{a}_2 \leqslant \mathfrak{a}_3 \leqslant \ldots \leqslant \mathfrak{a}_k.$$

If A contains no element of length j, then, since $A_j = A_{j+1} = \ldots = 0$, must

$$\mathfrak{a}_j = \mathfrak{a}_{j+1} = \ldots = 0.$$

Now let us suppose that B is an actual divisor of A, then

$$R < A < B \leqslant M,$$

so that every element

$$a_{1\lambda}\xi_1 + \ldots + a_{\lambda\lambda}\xi_\lambda$$

* Van der Waerden, *loc. cit.* p. 87.

of A_λ is contained in B, since $A_\lambda \leqslant A < B$; it is also contained in B_λ since it is at most of length λ. Thus $A_\lambda \leqslant B_\lambda$ and similarly

$$\mathfrak{a}_\lambda \leqslant \mathfrak{b}_\lambda \quad \ldots\ldots\ldots\ldots\ldots\ldots\ldots\ldots (40{\cdot}2).$$

Since B is an actual divisor of A, there must exist in B one or more elements which are not contained in A. Let $\beta = b_1\xi_1 + \ldots + b_i\xi_i$ be such an element whose length i is a minimum. If $\mathfrak{b}_i$ were not an actual divisor of $\mathfrak{a}_i$, then every b_j of $\mathfrak{b}_i$ would be contained in $\mathfrak{a}_i$ and

$$a = a_{1,i}\xi_1 + \ldots + a_{i-1,i}\xi_{i-1} + b_i\xi_i$$

would be an element of A_i and therefore of A also. Now if a is in B_i, so is $\gamma = \beta - a = (b_1 - a_{1,i})\,\xi_1 + \ldots + (b_{i-1} - a_{i-1,i})\,\xi_{i-1}$, but γ cannot be contained in A for otherwise β would be contained in A which is contrary to supposition. Hence γ is an element of B but not of A and is of length $\leqslant i-1$, which is also contrary to supposition; hence $\mathfrak{b}_i$ is an actual divisor of $\mathfrak{a}_i$. We have shown therefore that if $A < B$, then $\mathfrak{a}_i < \mathfrak{b}_i$ for some minimum value of i.

Suppose now $A^{(1)} < A^{(2)} < A^{(3)} < \ldots$, i.e. that we have a chain of modules, and suppose that the theorem of divisor chains holds for the ring R, then associated with $A^{(r)}$ are the ideals $\mathfrak{a}_1^{(r)} < \mathfrak{a}_2^{(r)} < \ldots < \mathfrak{a}_k^{(r)}$.

From equation (40·2) we have for every $\lambda = 1, 2, \ldots, k$ that

$$\mathfrak{a}_\lambda^{(1)} \leqslant \mathfrak{a}_\lambda^{(2)} \leqslant \mathfrak{a}_\lambda^{(3)} \leqslant \ldots.$$

But by hypothesis this last must be a finite chain. Let the last member be $\mathfrak{a}_\lambda^{(r_\lambda)}$ and let the maximum r_λ be r_0, so that $\mathfrak{a}_\lambda^{(r_0)} = \mathfrak{a}_\lambda^{(r_0+1)} = \ldots$ for all values of λ. But since $A^{(r_0)} < A^{(r_0+1)}$, a λ exists for which $\mathfrak{a}_\lambda^{(r_0)} < \mathfrak{a}_\lambda^{(r_0+1)}$ and we have a contradiction; therefore the chain $A^{(1)} < A^{(2)} < A^{(3)} < \ldots$ is a finite one. This theorem can be stated as follows: *If S is a finite super-ring of R and if every ideal in R is finite, then every R-module in S has a finite basis.*

Suppose that R is an actual sub-ring of F, and let a be an element of F but not of R. We can therefore form a super-ring of R containing all the elements in F of the type $r_0 + r_1 a + \ldots + r_m a^m$, where m is arbitrary. Here we have the R-modules

$$A_0 = (a^0) = (e) = R, \quad A_1 = (a^0, a^1), \quad A_2 = (a^0, a^1, a^2) \ldots (40{\cdot}3).$$

This chain is a divisor chain since $A_i \leqslant A_{i+1}$.

If for a certain m, $A_m = A_{m-1}$, then there must be an equation of the following type

$$a^m + r_{m-1}a^{m-1} + \ldots + r_0 = 0 \quad \ldots\ldots\ldots\ldots\ldots (40{\cdot}4),$$

with integral coefficients $r_0, \ldots, r_{m-1}$, and therefore every polynomial of degree $m+p$ in a is contained in A_{m-1}. In this case the super-ring has a finite module basis $\alpha^0, \alpha^1, \ldots, \alpha^{m-1}$. When α satisfies such an equation as (40·4) we say that it is entirely algebraic with respect to R, or shortly that it is **R-ENTIRE***.

From definition an R-entire element α defines a finite super-ring of R which we shall denote by R_α. We mean by this that R_α is finite with respect to R: we do not mean that R_α is itself a finite ring.

THEOREM 1. *Every element of a finite super-ring S of R is entire.*

Let S be a finite super-ring of R so that $R < S$ and let α be an arbitrary element of S. Let $\sigma_1, \ldots, \sigma_n$ form a finite module basis of S. Now since S is a ring, $\alpha\sigma_i$ belongs to S, so that

$$\alpha\sigma_i = r_{i1}\sigma_1 + r_{i2}\sigma_2 + \ldots + r_{in}\sigma_n.$$

Hence, using δ in the Kronecker sense, $|r_{ik} - \alpha\delta_k^i| = 0$ or $\alpha^n + \ldots = 0$; that is to say α is entire.

THEOREM 2. *If S is a super-ring of R and if every element of S is R-entire and if α from T, a super-ring of S, be S-entire, then α also is R-entire.*

Since α is S-entire, then $\alpha^m + \sigma_{m-1}\alpha^{m-1} + \ldots + \sigma_0 = 0$, where every σ_i is from S and is therefore R-entire. Hence α is also entire with respect to the super-ring $\bar{S}$ of R obtained by adjoining the elements $\sigma_0, \ldots, \sigma_{m-1}$ to R. Since every σ_i is R-entire, $\bar{S}$ is a finite super-ring of R; also α is S-entire, so that S_α and $\bar{S}_\alpha$ are finite super-rings of S and $\bar{S}$ respectively. $\bar{S}_\alpha$ is therefore a finite super-ring of R, so that α is R-entire by the last theorem. If every R-entire element of the quotient field P of R is an element of R, then we say that R is **ENTIRELY CLOSED.**

THEOREM 3. *If any element of R can be resolved into prime factors uniquely, then R is entirely closed.*

For, let α be an R-entire element of P, then since α is R-entire

$$\alpha^m + r_1\alpha^{m-1} + \ldots + r_m = 0 \qquad (r_i \text{ from } R).$$

Since α belongs to P, then $\alpha = \dfrac{t}{s}$, t and s being elements of R. Thus

$$t^m + r_1 t^{m-1}s + \ldots + r_m s^m = 0,$$

and s divides t^m and therefore divides t. Let $t = ss_1$, then $\alpha = \dfrac{ss_1}{s} = s_1$, an element of R itself. That is to say, R is entirely closed.

* Van der Waerden, *Moderne Algebra*, Teil 2, pp. 88–89, and Landau, *Zahlentheorie*, Bd. 3, p. 32.

§ 41. A Theorem of Artin and of van der Waerden.

Suppose that R is a commutative ring with a unit element and with no divisors of zero. We shall further suppose (1) that every ideal in this ring is finite and that therefore the theorem of divisor chains applies; (2) that R is entirely closed in its quotient field P; (3) that if R has a characteristic $p>0$, then is the radical-ring $R^{\frac{1}{p}}$ finite with respect to R. The third of these conditions is satisfied if R is a field which is either finite or perfect, or still more generally if R is a finite algebraic or a transcendental expansion of a finite or perfect field. If $R^{\frac{1}{p}}$ be finite with respect to R, so also is $R^{\frac{1}{p^e}}$ finite with respect to R. We shall now prove van der Waerden's Theorem*. *Let P be a quotient field of R, Σ a finite expansion of P, S the ring of all R-entire elements of Σ; then if conditions* (1), (2) *and* (3) *obtain for R, they also obtain for S.*

We shall first prove that S is entirely closed. This means that if σ from Σ is entire with respect to S, then σ is an element of S itself. Now if σ is entirely algebraic with respect to S, then $\sigma^m+s_1\sigma^{m-1}+\ldots+s_m=0$ and the s_i's are R-entire, and therefore σ is R-entire so that σ belongs to S. Hence S is entirely closed.

We must now show that S is a finite R-module, for the theorem of divisor chains would then hold for S. Also S would be finite with respect to R and so, following the isomorphism, $S^{\frac{1}{p}}$ would be finite with respect to $R^{\frac{1}{p}}$, and $R^{\frac{1}{p}}$ is finite with respect to R, so that $S^{\frac{1}{p}}$ would be finite with respect to R and hence with respect to S, since $R<S<S^{\frac{1}{p}}$; and hence condition (3) would be satisfied. We have therefore still to show that S is a finite R-module.

Now Σ is a finite and therefore an algebraic expansion of P. The most general finite expansion of P is obtained by an expansion Γ of the first sort followed by finding a radical field of Γ. If e be the exponent of Σ, i.e. $e=$ maximum of all the exponents of the elements of Σ, then $\Gamma\leqslant\Sigma\leqslant\Gamma^{\frac{1}{p^e}}$. Now let C be the ring of all the R-entire elements in Γ, then $C^{\frac{1}{p^e}}$ is the ring of all R-entire elements in $\Gamma^{\frac{1}{p^e}}$ since an element of $\Gamma^{\frac{1}{p^e}}$ is entire if, and only if, its p^eth power is entire, i.e. if its p^eth power is in C. Hence $C\leqslant S\leqslant C^{\frac{1}{p^e}}$.

* *Göttinger Nachrichten*, Heft 1, p. 26 (1926).

Now Γ is a finite expansion of the first sort of the field P, thus $\Gamma = P(\alpha)$. Every element γ of Γ and therefore every element of C can be written as

$$\gamma = \rho_0 + \rho_1 \alpha + \rho_2 \alpha^2 + \dots + \rho_{n-1} \alpha^{n-1},$$

where each ρ_i is from P. Since Γ is an expansion of the first sort, therefore the equation which defined α,

$$\alpha^n + \rho' \alpha^{n-1} + \dots + \rho^{(n)} = 0,$$

has n different roots. Let these be $\alpha_1, \alpha_2, \dots, \alpha_n$, so that we have n equations

$$\gamma_\nu = \rho_0 + \rho_1 \alpha_\nu + \rho_2 \alpha_\nu^2 + \dots + \rho_{n-1} \alpha_\nu^{n-1} \qquad (\nu = 1, 2, \dots, n),$$

and solving these for ρ_μ we have

$$\begin{aligned}\rho_\mu &= \frac{|1, \alpha, \alpha^2, \dots, \alpha^{\mu-1}, \gamma, \alpha^{\mu+1}, \dots, \alpha^{n-1}|_{1, 2, \dots, n}}{|1, \alpha, \dots\dots\dots\dots, \alpha^{n-1}|_{1, 2, \dots, n}} \\ &= \frac{|1, \dots, \alpha^{\mu-1}, \gamma, \alpha^{\mu+1}, \dots, \alpha^{n-1}|_{1, 2, \dots, n} \, |1, \alpha, \dots, \alpha^{n-1}|_{1, 2, \dots, n}}{(|1, \alpha, \dots\dots\dots, \alpha^{n-1}|_{1, 2, \dots, n})^2} \\ &= \frac{\tau_\mu}{D}, \text{ say.}\end{aligned}$$

Now D is the square of the product of the differences of the roots and is therefore rational. Hence D is an entire element of P, i.e. D belongs to R. Also $\tau_\mu = D\rho_\mu$, that is, it belongs to P; also it is entire since it is an integral function of the α's and is therefore also an element of R. Hence

$$\gamma = \frac{1}{D}(\tau_0 + \tau_1 \alpha + \dots + \tau_{n-1} \alpha^{n-1}).$$

Every element γ belonging to C can be represented in this manner, so that C is a module contained in the finite module $M = P\left(\frac{1}{D}, \frac{\alpha}{D}, \dots \frac{\alpha^{n-1}}{D}\right)$. Since M is a finite R-module, so is C, for $R \leqslant C \leqslant M$. Now C is finite with respect to R, and hence $C^{\frac{1}{p^e}}$ is finite with respect to $R^{\frac{1}{p^e}}$; also by isomorphism, $R^{\frac{1}{p^e}}$ is finite with respect to R, so that $C^{\frac{1}{p^e}}$ must be finite with respect to R; but $C \leqslant S \leqslant C^{\frac{1}{p^e}}$. Hence S is finite with respect to R and is an R-module. This completes the proof of van der Waerden's Theorem.

§ 42. The Finiteness Criterion of E. Noether*.

Let P be a given field and let $P[x_1, \dots, x_n]$ be the polynomial ring of rational integral polynomials in the $x_1, \dots, x_n$ with coefficients belonging to P. Further, let J be a division ring whose elements are

* *Göttinger Nachrichten* (1926), p. 31.

P polynomials such that $P < J < P[x_1, \ldots, x_n]$. J is finite with respect to P whenever there is contained in J a finite number of elements $I_1, I_2, \ldots, I_h$, so that every element of J can be expressed as a rational integral function, with coefficients from P, of this basis $I_1, \ldots, I_h$, in which case

$$P < J \leqslant P[I_1, \ldots, I_h] \leqslant P[x_1, \ldots, x_n].$$

We shall suppose also that if P has the characteristic p, then $P^{\frac{1}{p}}$ is finite with respect to P.

E. NOETHER'S THEOREM. *A ring J of polynomials in $x_1, \ldots, x_n$ with coefficients from P, which has no divisors of zero, is finite with respect to P if, and only if, there exists within J a sub-ring R which is finite with respect to P, such that every element of J is R-entire.*

This condition is necessary; for if J is finite with respect to P, then we can put $R=J$. It is also shown to be sufficient as follows. Suppose that K is the quotient field of R and that L is the quotient field of J. This is possible since the rings have no divisors of zero. Hence

$$P < K < L < P(x),$$

where $P(x)$ is the field of all rational functions of x with coefficients from P. We now make use of the corollary given at the end of § 36. It can be demonstrated readily that L is an algebraic expansion of K, and hence L is a finite expansion of K, where, if P be a $P^{(p)}$ with $p > 0$, we understand the most general expansion of the first or second sort. Call S the ring of R-entire elements of L. Since therefore every element of J is R-entire, J is a sub-ring of S, i.e. $J < S < L$.

Suppose now that R be entirely closed. Since R is finite with respect to P, then $R = P[g_1(x), \ldots, g_r(x)]$, where $g_i(x)$ is a polynomial in $x_1, \ldots, x_n$ with coefficients from P and belongs to R and is therefore an element of J. Since R is a finite ring the theorem of divisor chains applies and we can also use van der Waerden's Theorem. R is a finite ring and K is its quotient field. L is a finite expansion of K; S is the ring of R-entire elements of L. The theorem of divisor chains for R-modules therefore holds for S. J is an R-module and has therefore a finite module basis $h_1(x), \ldots, h_s(x)$, i.e. every $f(x)$ of J can be written as $f(x) = A_1 h_1(x) + \ldots + A_s h_s(x)$, where the A's belong to R. The A's are therefore polynomials in the g's so that the g's and h's together form a finite module basis of K.

If R be not entirely closed, then we must first construct a sub-ring T of R which is finite with respect to P and which is entirely closed and with respect to which R is entire. We can proceed with the above proof, using T in place of R. We refer the reader to E. Noether's paper for this case.

§43. Application of E. Noether's Theorem to Modular Covariants*.

Consider the group Γ of all homogeneous linear substitutions with coefficients from a Galois field, and let $A_1, A_2, \ldots, A_h$ be the different substitutions where A_1 is the identity substitution. Now let the transformation A_k transform the set of variables $x_1^{(1)}, \ldots, x_m^{(1)}$ to the set $x_1^{(k)}, \ldots, x_m^{(k)}$. Consider also the resolvent of Galois,

$$\phi(z, u) = \prod_{k=1}^{h} (z - u_1 x_1^{(k)} - u_2 x_2^{(k)} - \ldots - u_m x_m^{(k)})$$

$$= z^h + \Sigma\, U_{a a_1 \ldots a_m}(x)\, z^a u_1^{a_1} \ldots u_m^{a_m},$$

where $a < h$ and $a + a_1 + \ldots + a_m = h$.

Let $\theta_k = \sum_{i=1}^{m} u_i x_i^{(k)}$. We can choose the u_i's so that $\theta_k \neq \theta_j$ unless $k = j$, so that any symmetrical function of $\theta_1, \ldots, \theta_h$ must be an absolute covariant. $\phi(z, u)$ is such a function and hence the coefficients $U_{a a_1 \ldots a_m}(x)$ are absolute covariants.

Let J be the ring of all absolute covariants and R the ring obtained from all the elements $U_{a a_1 \ldots a_m}(x)$ so that $R \leqslant J$. If we can show that every element I of J is R-entire, then from the theorem of the previous paragraph, J will have a finite basis provided as before that $P^{\frac{1}{p}}$ is finite with respect to P.

Now if we put $z = u_1 x_1^{(1)} + \ldots + u_m x_m^{(1)}$, then

$$(u_1 x_1^{(1)} + \ldots + u_m x_m^{(1)})^h + \Sigma\, U_{a a_1 \ldots a_m}(x)\,(u_1 x_1^{(1)} + \ldots + u_m x_m^{(1)})^a u_1^{a_1} \ldots u_m^{a_m} \equiv 0$$

for all u_i's. Suppose then that $u_j = \delta_j^i$ (using δ in the Kronecker sense), then

$$x_i^h + \sum_a x_i^a\, U_{a 0 \ldots 0\, a_i\, 0 \ldots 0}(x) = 0,$$

where $a + a_i = h$ and $a < h$. This last equation shows that x_i is entirely algebraic with respect to a portion of R and therefore with respect to R itself. It follows that every polynomial in the x's and therefore every element of J is entire with respect to R.

* *Göttinger Nachrichten* (1926), p. 33.

If the field P is the $GF[p^n]$, then $P^{\frac{1}{p}}$ is finite with respect to P and all the conditions are satisfied and the theorem is proved for the case of universal covariants. The extension is made to covariants involving coefficients of the forms by regarding the coefficients as additional variables.

The proof also holds for relative covariants as well as for absolute, since the factor appearing also belongs to P. The proofs also hold if the x's and a's are unknowns of the $GF[p^n]$, so that we have proved the finiteness of covariants for every case in the theory of modular covariants.

APPENDIX I

Dickson in his *History of the Theory of Numbers*, vol. 3, chap. 19, has given a summary of all the papers printed before 1922 on the subject of Modular Invariants. The present Appendix is intended to bring Dickson's work up to date.

Moore* gave a twofold generalisation of Fermat's Theorem and showed that the determinant

$$\begin{vmatrix} x_1, & x_2, & \dots, & x_m \\ x_1^{p^n}, & x_2^{p^n}, & \dots, & x_m^{p^n} \\ \vdots & & & \\ x_1^{p^{n(m-1)}}, & & \dots\dots, & x_m^{p^{n(m-1)}} \end{vmatrix}$$

is congruent to the product of all distinct non-zero linear forms with coefficients in the $GF[p^n]$. This determinant is none other than the universal covariant L_m which we discussed in § 12.

Hazlett† discussed the relationship between the theory of modular invariants and the theory of projective invariants. She developed the use of symbolical notation and showed the important difference between isobaric and pseudo-isobaric covariants. It is shown that all formal binary invariants admit of symbolical representation. Methods of finding the symbolical representation of an invariant are given. There is an error in this paper which Miss Hazlett ‡ later corrected.

Feldstein§ gave the full system of universal covariants of the m-ary group of transformations whose coefficients are the positive integral residues mod t, where $t = p^k$; viz.

$$L_m^{p^{k-1}}, \quad Q_{m,s}^{p^{k-1}}, \quad p^j L_m^{ap^{k-j-1}} \prod_{s=1}^{n-1} Q_{m,s}^{b_s p^{k-j-1}},$$

where $s = 1, \dots, m = 1$; $j = 1, \dots, k-1$; and where the a and b_s s range over $0, 1, \dots, p-1$, but may not all be zero. The above results were given in an earlier paper for the case where $k = 2$ by Turner‖.

* *Bulletin of the American Mathematical Society*, vol. 2, p. 189 (1895–96).

† *Transactions of the American Mathematical Society*, vol. 24, pp. 286–311 (1922).

‡ *Ibid.* vol. 30, p. 855 (1930).

§ *Ibid.* vol. 25, pp. 223–238 (1923).

‖ *Ibid.* vol. 24, pp. 129–134 (1922).

Glenn* gave a full system of the formal covariants of two binary quadratics in the $GF[2]$. A new method of obtaining covariants is described which depends upon the "appropriate selection" of a primary quantic. If the selection be not "appropriate" then no covariants may result, so that this method is of no great value until we have some more definite method of selecting the primary quantic.

Gouwens† extended the results of a paper by Mrs Ballantine‡ from 2 to m variables. It is proved that every invariant of the group of transformations with determinant congruent to unity modulo $\Pi = p_1^{\lambda_1} p_2^{\lambda_2} \dots p_r^{\lambda_r}$ is a sum of invariants, each of which is expressible as a product of $k_i = \frac{\Pi}{p_i^{\lambda_i}}$ by an invariant of the group H_i of transformations with determinant congruent to unity mod $p_i^{\lambda_i}$. Conversely every such product is an invariant.

Williams§ treated full systems of formal modular protomorphs of binary forms. Elliott|| gave the algebraic protomorphs of the transformation $x = \bar{x} + t\bar{y}$, $y = \bar{y}$, viz.

$$S_1 = a_0, \quad S_2 = a_0 a_2 - a_1^2, \quad S_3 = a_0^2 a_3 - 3a_0 a_1 a_2 + 2a_1^3, \; \dots.$$

It is shown that the seminvariants $S_i\,(i = 1, 2, \dots, l)$ and $a_1^p - a_0^{p-1} a_1$ form a full system of protomorphs of the binary l-ic mod p, where p is such a prime that $\binom{l}{j} \not\equiv 0 \bmod p$ $(j = 1, 2, \dots, l-1)$. The theorem is also given for the case of several binary l-ics.

Hazlett¶ extended the results of a paper by Williams**. Let $f(x_1, \dots, x_m)$ be any homogeneous polynomial in n variables of order l, and let F_1 be any homogeneous polynomial in the values of f as $(x_1, \dots, x_m)$ range over the real points of the field. Let F_i range over all the conjugates of F_1, under transformations of the group $\Gamma_1^{(m)}$, which are incongruent in the field. If $\chi = \lambda(p^n - 1)/l$ be an integer, where λ is some fixed positive integer, then any symmetric function of the χth powers of the F_i is a formal invariant of f under the group $\Gamma_1^{(m)}$ with

* *Bulletin of the American Mathematical Society*, vol. 30, pp. 131–139 (1924).

† *Transactions of the American Mathematical Society*, vol. 26, pp. 435–440 (1924).

‡ *American Journal of Mathematics*, vol. 45, pp. 286–293 (1923).

§ *Transactions of the American Mathematical Society*, vol. 28, pp. 183–197 (1926).

|| *Algebra of Quantics*, pp. 212–215.

¶ *Journal de Mathématique*, Ser. 9, vol. 9, pp. 327–332 (1930).

** *Ibid.* vol. 4, pp. 169–192 (1925).

respect to the $GF[p^n]$. This result is similar to Dickson's* method of obtaining formal covariants.

E. Noether† proved the finiteness of modular covariants by using the Theory of Fields.

APPENDIX II

In this Appendix we give a list of papers on the subject of Modular Invariants. The contents of these are summarised either in Appendix I or by Dickson in his *History of the Theory of Numbers*. We shall use the following abbreviations:

T.A.M.S. = *Transactions of the American Mathematical Society.*
B.A.M.S. = *Bulletin of the American Mathematical Society.*
A.J.M. = *American Journal of Mathematics.*
P.L.M.S. = *Proceedings of the London Mathematical Society.*
Q.J.M. = *Quarterly Journal of Mathematics.*
A.M. = *Annals of Mathematics.*
J.M. = *Journal de Mathématique.*
P.N.A.S. = *Proceedings of the National Academy of Sciences.*

Author	Periodical	Volume	Pages	Year	Ref. No.
Hurwitz	*Archiv d. Math. u. Phys.* (3)	5	17–27	1903	1
Dickson	*T.A.M.S.*	8	205–232	1907	2
„	*P.L.M.S.* (2)	5	301–324	1907	3
„	*A.J.M.*	30	263–281	1908	4
„	*T.A.M.S.*	10	123–158	1909	5
„	*Q.J.M.*	40	349–366	1909	6
„	*P.L.M.S.* (2)	7	430–444	1909	7
„	*A.J.M.*	31	103–146	1909	8
„	*A.J.M.*	31	337–354	1909	9
„	*T.A.M.S.*	12	1–18	1911	10
„	*Q.J.M.*	42	158–161	1911	11
„	*T.A.M.S.*	12	75–98	1911	12
„	*B.A.M.S.*	19	456–457	1912–13	13
„	*A.J.M.*	33	175–192	1913	14
„	*B.A.M.S.*	20	132–134	1913–14	15
„	*T.A.M.S.*	14	299–310	1913	16
„	*A.M.* (2)	15	114–117	1913–14	17
„	*Madison Colloquium Lectures*			1914	18

* *Transactions of the American Mathematical Society*, vol. 15, pp. 497–503 (1914).

† *Göttinger Nachrichten* (1926), pp. 28–35.

Author	Periodical	Volume	Pages	Year	Ref. No.
Dresden	*B.A.M.S.*	20	116–119	1913–14	19
Glenn	*B.A.M.S.*	21	464–470	1914–15	20
Dickson	*Q.J.M.*	45	373–384	1914	21
Krathwohl	*A.J.M.*	36	449–460	1914	22
Wiley	*T.A.M.S.*	15	431–438	1914	23
Dickson	*T.A.M.S.*	15	497–503	1914	24
„	*B.A.M.S.*	21	174–179	1914–15	25
„	*A.J.M.*	37	107–116	1915	26
„	*P.N.A.S.*	1	1–4	1915	27
„	*A.J.M.*	37	337–354	1915	28
„	*A.J.M.*	37	355–358	1915	29
McAtee	*A.J.M.*	41	225–242	1919	30
Hazlett	*A.J.M.*	43	189–198	1921	31
„	*T.A.M.S.*	21	247–254	1920	32
„	*T.A.M.S.*	22	144–157	1921	33
Sanderson	*T.A.M.S.*	14	489–500	1913	34
Glenn	*B.A.M.S.*	21	167–173	1914–15	35
„	*A.J.M.*	37	73–78	1915	36
„	*T.A.M.S.*	17	545–556	1916	37
„	*T.A.M.S.*	18	460–462	1917	38
„	*A.M.* (2)	19	201–206	1917–18	39
„	*T.A.M.S.*	19	109–118	1918	40
„	*T.A.M.S.*	20	154–168	1919	41
„	*T.A.M.S.*	21	285–312	1920	42
„	*P.N.A.S.*	5	107–110	1919	43
Williams	*T.A.M.S.*	22	56–79	1921	44
Hazlett	*T.A.M.S.*	24	286–311	1922	45
Turner	*T.A.M.S.*	24	129–134	1922	46
Feldstein	*T.A.M.S.*	25	223–238	1923	47
Glenn	*B.A.M.S.*	30	131–139	1924	48
Ballantine	*A.J.M.*	45	286–293	1923	49
Gouwens	*T.A.M.S.*	26	435–440	1924	50
Williams	*T.A.M.S.*	28	183–197	1926	51
„	*J.M.* (9)	4	169–192	1925	52
Hazlett	*J.M.* (9)	9	327–332	1930	53
Noether	*Göttinger Nachrichten*		28–35	1926	54
Dickson	*History of the Theory of Numbers*, vol. 3, chap. 19			1923	55

APPENDIX III

We tabulate here the papers in which the modular covariants of an m-ary l-ic are considered for particular values of m and l. The numbers of the papers refer to those of Appendix II. $l=1+2$ denotes that the simultaneous covariants of a linear and a quadratic form are treated.

m	l	Papers
2	1	35, 42
2	2	2, 3, 4, 5, 8, 9, 16, 18, 19, 24, 25, 30, 31, 33, 34, 36, 41, 42, 43, 45, 52
2	3	2, 18, 36, 37, 40, 43, 45, 52
2	4	17, 18, 37, 42, 52
2	5	37, 52
2	6	37
2	7	37
2	m	18, 44
2	1+1	48
2	1+2	9, 24
2	2+2	6, 8, 9, 48, 51
2	1+2+3	21
3	2	2, 3, 4, 13, 18, 24, 25
3	3	26
3	4	28
3	1+1	29
3	2+2	6
4	2	3, 25
5	2	3
6	2	3
m	2	3, 4
m	2+2	7

INDEX

Printed in the United States
By Bookmasters